TIME-BASED COMPETITION

THE NEXT BATTLEGROUND IN
AMERICAN MANUFACTURING

**The BUSINESS ONE IRWIN/APICS Series in
Production Management**

**Supported by the American Production
and Inventory Control Society**

**OTHER BOOKS PUBLISHED IN THE BUSINESS ONE IRWIN SERIES
IN PRODUCTION MANAGEMENT**

Attaining Manufacturing Excellence   Robert W. Hall
Bills of Materials   Hal Mather
Production Activity Control   Steven A. Melnyk and Phillip L. Carter
Manufacturing Planning and Control Systems,   Second Edition
Thomas E. Vollmann, William Lee Berry, D. Clay Whybark
The Spirit of Manufacturing Excellence   Ernest Huge
Strategic Manufacturing, Dynamic New Directions for the 1990's
Patricia E. Moody
Total Quality: An Executive's Guide for the 1990s   The Ernst & Young
Quality Improvement Consulting Group

The BUSINESS ONE IRWIN/APICS Series in
Production Management

APICS ADVISORY BOARD

L. James Burlingame
Eliyahu M. Goldratt
Robert W. Hall
Ed Heard
Ernest C. Huge
Henry H. Jordan
George W. Plossl
Richard Schonberger
Thomas E. Vollmann
D. Clay Whybark

TIME-BASED
COMPETITION
THE NEXT BATTLEGROUND IN
AMERICAN MANUFACTURING

The BUSINESS ONE IRWIN
APICS Series in Production Management

Joseph D. Blackburn
Editor

BUSINESS ONE IRWIN
Homewood, Illinois 60430

To Ian, Jessica, and Brent,
who define value-added time for me

Project editor: Lynne Basler
Production manager: Carma W. Fazio
Artist: Regina Megacz
Compositor: Eastern Graphics
Typeface: 11/13 Century Schoolbook
Printer: R. R. Donnelley & Sons Company

Library of Congress Cataloging-in-Publication Data

Time-based competition : the next battleground in American
  manufacturing / Joseph Blackburn, editor.
     p.   cm. — (APICS series)
  Includes index.
  ISBN 1-55623-321-3
  1. United States—Manufactures—Management. 2. Just-in-time
systems—United States. 3. Competition—United States.
I. Blackburn, Joseph. II. Series: BUSINESS ONE IRWIN/APICS
series in production management.
HD9725.T57 1991
658.5—dc20   90–3471
                                       CIP

Printed in the United States of America
  3 4 5 6 7 8 9 0 DOC 7 6 5 4 3 2 1

PREFACE

Many manufacturing executives recognize the significance of response time as a competitive weapon but are struggling to achieve world-class speed. *Time-Based Competition* is for them. It explains how other U.S. firms have dealt with the problems of time-compressing different segments of the value-delivery chain—new product development, manufacturing, and distribution. With the exception of fine wines and art, few industries seem to be immune from time competition. In every industry we have studied, at least one firm leads in terms of speed of response to the market, and, most significantly, these time-competitive firms are emerging as the market leaders.

Not all managers, however, recognize how time-based competition already pervades the industries in which they compete. For example, in industries such as machine tools and plastic injection molds, sizable numbers of firms do not view time as strategically important. Management in these types of industries has become accustomed to long lead times and does not sense the winds of change sweeping through their industries. *Time-Based Competition* was written for this group, too. For them, time is a critical dimension of competition that they could learn about too late—when their competitors use the weapon against them.

This book is also for my academic colleagues who are continually searching for new research ideas in the management field of operations. Although the research literature on Just-in-Time (JIT) in manufacturing is extensive, much more research is needed on time-compression of activities outside the factory walls—in new product development, in customer service, and in the white-collar environment of the office. Focusing on time in the value-delivery chain of an organization raises many challenging research issues, and this book raises questions that will head researchers toward those issues.

My colleagues may wonder how this long-time operations researcher can write a book without a single equation or theorem. The study of the strategic consequences of time does not yet lend itself to mathematical models. Most of the research to date has been empirical and has relied heavily on anecdotal evidence. This should change, however, as we learn to build better models of functions such as new product development. So I do not apologize for the absence of equations in this work. Instead I would point to the fact that the most profound book about time written for the general reader, Stephen Hawkings's *A Brief History of Time*, contains only a single equation: Einstein's $E = mc^2$.

The event that produced much of the material for this book was a conference on time-based competition that I organized at the Owen Graduate School of Management of Vanderbilt University in December 1988. Although this book was spawned by the discourse at that conference, the conference itself actually was the culmination of two years' work by a research group, the Operations Roundtable, started by the Owen Graduate School of Management. The gulf that separates academics and practitioners is quite wide and often seems unspannable. The Operations Roundtable was formed as a research partnership to bring the two groups together and focus on real problems. Through the roundtable I not only became convinced of the strategic importance of time compression, but I also came to question some of my fundamental research models.

How the Operations Roundtable and I became focused on the issue of time is important because it illustrates the value of academic-industry cooperative ventures. The first meeting of the roundtable in 1986 produced cultural whiplash among both the academics and the manufacturing executives when we began trying to set a research agenda. During the first few hours, several faculty members talked about their research interests. I soon observed that the yawn rate in the room was reaching alarming levels; the executives were leaving the room at frequent intervals to check on phone calls back at the office. Ennui had replaced the excitement with which the day had begun.

I knew something was wrong, but wasn't quite sure what. Finally, during a break, one executive came to me and said, "Why don't you ask us about our problems?" Our (academicians') mistake was a failure to recognize the difference between what Bill Maxwell, a Cornell professor, calls "manufacturing problems and manufactured problems." The executives were not particularly interested in our "manufactured problems"—another scheduling algorithm, a forecasting technique, or an inventory-distribution model. They not only didn't wonder if there exists an "optimal solution," they didn't care. They had real problems and needed real answers. When we finally got around to asking the managers about their problems, we got an earful. Typical questions were, "How do we get new products developed faster? How do we create a climate for change in the organization? What things do we measure? How do we give our customers better service?"

From the lively discussion that followed, time emerged as the research theme that unified the roundtable members. We had George Stalk of the Boston Consulting Group to thank for sparking that interest. In our initial meetings, George's explanations of how Japanese firms were using time-based competition to become a dominant force in manufacturing captured our attention and focused the group's interest on the single research topic of time. Each firm in the roundtable recognized the importance of time as a unifying concept and strongly supported a research program to learn how they could time compress their own operations. This interest led to the conference in 1988.

For me, George Stalk's theories brought a revelation. It all suddenly made sense. I realized, after reading all about JIT and zero inventory and Japanese management, that inventory reduction was only a small part of the story. Inventory reduction is a benefit of JIT, but not its *raison d'etre*. Time is the elemental concept that transfers to other functions within the organization.

There is something simple and elegant about focusing on time. It is a powerful concept because making a process faster and more responsive also reduces cost, improves quality, and lowers inventory. In addition, time-compression supports the currently popular admonition to "get closer to the customer." A firm's pursuit of speed, however, presents its management with a continuing set of challenges, so the journey, although endless, is endlessly rewarding.

Joe Blackburn, Editor

ACKNOWLEDGMENTS

Although I should thank hundreds of people for their support in the writing of this book, my deepest thanks go to those who spoke at the Time-Based Competition Conference, the event that spawned this book. The thoughts of our keynote speakers, George Stalk of the Boston Consulting Group and Fred Smith, the CEO of Federal Express, are central contributions to the book. Important chapters were contributed by Dan Ciampa of Rath & Strong, John Bailey of Honeywell, Andy Vecchione of Milliken, Mohan Kharbanda of Xerox, Roger Schmenner of Indiana University, Bob Millen of Northeastern University, and Stephen Hamilton of Hewlett-Packard. Others who made significant contributions to the conference were Tom Caste of Scott Fetzer, Ted Schlie of Lehigh University, Ed Parrish of Coopers & Lybrand, Bob Badelt of Northern Telecom, and Professors Bezalel Gavish, Tom Dibble, and Paul Makens of Vanderbilt University.

I remain deeply indebted to the member firms of the Operations Roundtable who supported the conference and the research that led to this book; special thanks are due to Bill Cawthon who helped found the roundtable before his retirement from Northern Telecom. (The roundtable firms are Northern Telecom, Coopers & Lybrand, Martin-Marietta Energy Systems, Acme Boot Company, France Division of Scott Fetzer, Fleetguard Division of Cummins Engine Company, Rath & Strong, Support Systems International, Ingram Industries, and Andersen Consulting.) Additional research support was provided by The Dean's Fund for Research of The Owen Graduate School of Management, Vanderbilt University.

The inspiration for the book came from two people—Bruce Henderson and George Stalk. All of us at the Owen School are grateful to Bruce for choosing to return to his alma mater upon retirement from the Boston Consulting Group. Most of what today's business students learn about strategy and competition is due to Bruce Henderson, and the insights into time competition that I have gained from conversations with (and interrrogations by) Bruce have been invaluable to me. But I am especially grateful because Bruce introduced me to his BCG colleague, George Stalk, on my first trip to Japan, and this has led to George's frequent trips to Nashville to share his ideas with the Operations Roundtable. Bruce Henderson defined the field of strategy as we know it, and within that field, George Stalk has demonstrated the power of time-focused strategies. Both men have shaped my views on time-based competition, although they are certainly not responsible for flaws in the logic of my conclusions.

Of my academic colleagues with whom I have exchanged ideas and who have refined my thinking about time competition, I would like to single out Bob Millen of Northeastern University. For many years, Bob has been my friend and research alter ego. He has helped me attain that clarity of vision that comes from discussing Japanese management techniques over many a Kirin and Sapporo.

I would like to thank all the graduate students who have worked on time-compression research projects—Mark Springer, Randy Mead, Scott Barbour, Mike Siebenmorgen, Jon Whaling, Alex Smith, Joe Herman, Susan Peppers, and Mary Ann Bilbro.

Finally, I would also like to thank those who have worked so hard with the editing and typing of the book—Norman Moore, Tom Plath, Rita Carswell, and Dan Vandermuelen. May Woods juggled my schedule and kept the associate dean's office running so that I could find the time to work on the book.

CONTENTS

PART 1

STRATEGIC OVERVIEW

CHAPTER 1

THE TIME FACTOR

Joseph D. Blackburn

As American firms struggle to regain an advantage over foreign competition, most neglect their most potent weapon: time. Geographically located in the midst of the world's most lucrative market, U.S. firms clearly hold the advantage needed to excel at responsiveness to the customer. Many U.S. firms, however, fail to exploit this advantage and, instead, find themselves observing while Japanese firms use the element of time to sustain their competitive edge. Consequently, the quickness of Japanese firms overcomes the sole natural advantage held by U.S. firms.

For the past decade, U.S. firms have assumed a follow-the-leader role in manufacturing strategy, with Japan in the enviable role of leader. At the hint of a new management trend emerging from Japan, legions of U.S. executives and academics boarded 747s to Tokyo, toured plants to learn new practices, and returned home to try to emulate the new practices. In business, however, the game of follow the leader is usually a losing game. When chasing a moving target a pursuant firm always lags a step or two behind the leader, and the leader tends to grab the lion's share of the profits.

Nevertheless, this book is not another prophesy of doomsday for U.S. industry because of Japan's prowess in manufacturing. On the contrary, American firms have an opportunity to change the rules of the game by focusing on time. In the current autoclave of world-class competition, the speed with which a firm can deliver a product or service to market has emerged as the dominant competitive factor. Some Japanese firms are particularly vulnerable to U.S. industrial advances because they have not yet adopted time-based strategies. Even those Japanese firms that practice time compression in manufacturing are at an enormous disadvantage in American markets because of the time required to move their products across the Pacific. Exploiting the time advantage is a way that U.S. firms can tilt the playing field in their favor and bring the customers sliding in their direction.

Foreign competition compels U.S. firms to reorient their corporate strategies toward time. Roy Merrills, president of Northern Telecom Incorporated, demonstrates this when he says that "all the things that were vital to our long-term competitiveness had one thing in common: time . . . emphasizing time instead of money means rethinking every aspect of the business."[1] Quality and price are still important because today's discriminating customer demands world-class quality at a competitive price. When all the leading firms in an industry have achieved high levels of quality, a focus on quality alone will not attract new customers. A faster response time must complement quality. According to Dean Cassell, vice president for product integrity at Grumman Corporation, "Speed and quality are not a trade-off. Speed is a component of quality—one of the things we must deliver to satisfy customers."[2]

On the other hand, firms that follow a conventional cost-based strategy are doomed to a shrinking market share in today's environment. In fact, such firms find that, after struggling to drive costs down and quality up, their domestic competition continues to mount, and, unfortunately, the offshore competition remains ahead. After exhorting the troops to wage war on costs and defects, top management finds, to its dismay, that the company has chased a moving target and lost the war.

---

[1]Roy Merrills, "How Northern Telecom Competes on Time," *Harvard Business Review*, July–August 1989, pp. 108–14.

[2]George H. Labovitz, "Speed on the Cycle Helps Companies Win the Race," *The Wall Street Journal*, October 30, 1989.

Given a choice between products (or services) of high quality and similar cost, customers gravitate toward the one that is delivered the fastest. Response time is the salient factor. Confederate General Nathan Bedford Forrest was more than a century ahead of his time with his battle plan to "git thar the fustest with the mostest." Today's firms, however, might substitute "bestest" for "mostest."

CUSTOMER DEMAND AND RESPONSE TIME

Cozying Up to the Customer

Today's—and tomorrow's—customers want it all: price, quality, and timely delivery. Marketing strategists advise firms to get closer to the customer—that is, learn what the customer needs and meet that need as quickly as possible. Because customers desire instant gratification for many products and services, reducing the time required to reach the customer must play a major role in a firm's competitive strategy. Firms must shrink the time from "conception to consumption"[3]—in other words, from the time consumers conceive the need for a product or service to the time that need is met: from seed to feed.

Procter & Gamble has long been the world leader in the marketing of packaged consumer products. P&G's goal has always been simple: each product should be number one in market share in its product category. In the mid-1980s, however, this dominance was challenged because consumers perceived P&G as being too slow in responding to the customer. Product managers complained that it took a year to get a simple change in a package design. P&G's vaunted brand management system was in danger of losing sight of the marketplace.[4]

In 1987, John Smale, the CEO at Procter and Gamble, moved to make the firm a time-based competitor by creating a new system of category management. In this system, the cate-

---

[3]Stanley Davis, *Future Perfect* (Reading, Mass: Addison-Wesley, 1987).
[4]Brian Dumaine, "P&G Rewrites the Marketing Rules," *Fortune*, November 6, 1989.

gory manager became the coach of a team that worked to guide the launching or repositioning of a product. Each team consisted of the brand manager and managers from manufacturing, sales, and finance—plus a new player, the product supply manager, who coordinates manufacturing, engineering, purchasing, and distribution. The product supply manager, moreover, became the time manager of the team, responsible for getting the product to market as quickly as possible.

Minor product design changes at P&G used to take up to a year or more. Faced with the need for an easier-to-use cap for its liquid dishwashing detergent Cascade, management turned the project over to a team. The process worked: the cap went into production, defect-free, in nine months—double the speed of the former conventional system.[5]

Procter & Gamble also worked with retailers to take time out of the distribution link in the chain. Joint projects on time compression and coordination of activities with major retailers such as Kroger and Wal-Mart moved P&G even closer to the customer and, at the same time, strengthened the firm's competitive position as the dominant supplier to these retailers. By focusing on time and the consumer, P&G successfully restructured its production and distribution systems and reasserted its dominance in many product segments. The company claimed the number one position in 22 product categories in 1989, up from 17 in 1985. Consequently, its earnings for 1989 increased 18 percent.[6]

Firms that compete globally need a customer focus if they are to thrive. Thomas Arenberg of Arthur Andersen, however, claims that "only 5 to 10 percent of the firms are doing the work it takes to get there"[7] (i.e., to a customer focus). David Kearns, the chief executive officer at Xerox, states that "customer satisfaction is the number one thing that drives market share."[8] A time-based strategy seeks to provide the quickest possible response to customer demands.

---

[5]Ibid.

[6]Ibid.

[7]Amanda Bennett and Carol Hymowitz, "For Customers, More than Lip Service," *The Wall Street Journal*, October 6, 1989, p. B1.

[8]Ibid.

What's New about Time?

When first hearing about time compression or time-based management, many managers take on a quizzical, skeptical look and say something like, "That's nothing new, is it? Haven't we had efficiency experts around for years? We've been through all that old stuff with stopwatches, clipboards, and time-and-motion study."

Time-based competition is not a resurrection of industrial engineering. It is more than a replacement of the industrial engineer in crepe-soled shoes by a consultant in pinstripes and wing tips. The efficiency experts focused on the "best way to do things"—they took a microscope to the workplace, broke a job down into its parts, and tried to ensure that the human labor in a task was performed at utmost efficiency. Focusing on time is not an outgrowth of the efficiency era of the 1930s.

Time-based competitors focus on the bigger picture, on the entire value-delivery system. They attempt to transform an entire organization into one focused on the total time required to deliver a product or service. Their goal is not to devise the best way to perform a task, but rather to either eliminate the task altogether or perform it in parallel with other tasks so that overall system response time is reduced. Becoming a time-based competitor requires making revolutionary changes in the ways that processes are organized.

Time Is Money?

The old saw "time is money" needs to be sharpened. "Time *instead* of money" might be more accurate. Money means costs, and most firms still are driven by cost-based strategies. Cost reduction dominated strategic thinking in the 1960s, and many of today's accounting and performance measurement systems were installed during that period of U.S. hegemony in the global economy. Firms that have pursued cost-cutting strategies while neglecting time are defenseless against a time-focused competitor with equivalent costs.

"Convert time into money" is an even better credo to underpin the goals of the time-focused competitor. To create real and sustainable market advantages, the time-focused competitor concentrates not on cutting cost, but on cutting time from all operations and, thus, making the firm more responsive to customer demand. This strategy creates heightened customer satisfaction that leads to increased sales and greater profits. In this way, time is converted to money.

Time reduction provides an important leveraging of profits that is not obtained with cost-reduction strategies. Firms find that by removing time from their operations, costs are reduced as well with no additional effort. In his pioneering *Study of the Toyota Production System*, Shigeo Shingo stated the reason. "Construct a production system that can respond without wastefulness to market change and that, moreover, by its very nature reduces costs."[9] More importantly, quality improvements also tend to accompany time compression; everything gets better.

The direct, potent relationship between time compression and cost reduction is clearly demonstrated in recent research by Marvin Lieberman of Stanford University. In a study of inventory reduction and productivity among Japanese auto manufacturers, Lieberman noted a strong relationship between throughput time and labor productivity.[10] The 1973 data for Japan's automakers shown in Exhibit 1 reveal that Toyota had the fastest throughput time and the highest labor productivity; Fuji, on the other hand, had the slowest throughput time, the lowest productivity, and the highest labor cost. These correlations and the positions of the other automakers suggest that there was an approximately linear functional relation between throughput time and productivity in 1973.

In a longitudinal analysis, Lieberman also observed that all world automakers decreased throughput time during the 1970s and 1980s. With the exception of Fuji, most firms also made significant progress in productivity. The 1987 data presented in Exhibit 2 show that lower throughput time was accompanied by

---

[9] Shigeo Shingo, *A Revolution in Manufacturing: The SMED System* (Cambridge, Mass.: Productivity, Inc., 1985), page xv.

[10] Marvin Lieberman, "Inventory Reduction and Productivity Growth, A Study of Japanese Automobile Producers," Chapter 21 in *Manufacturing Strategy*, John Ettlie, M. Burstein, and A. Feigenbaum, eds. (Boston: Kluwer Academic Publishers, 1990).

EXHIBIT 1
1973 Productivity versus Throughput Time

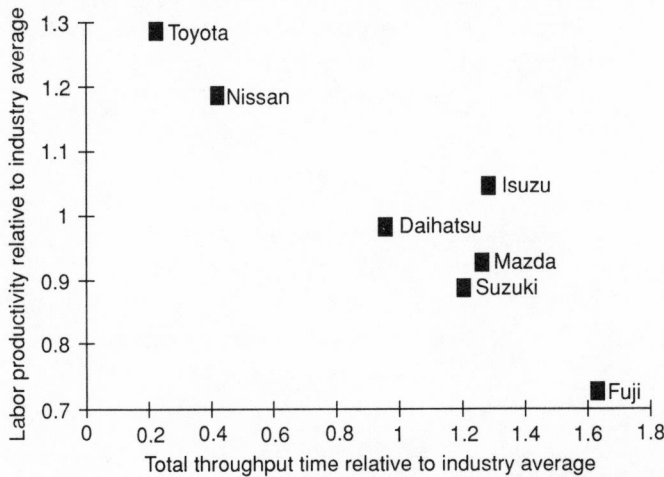

Source: Marvin Lieberman, "Inventory Reduction and Productivity Growth, A Study of Japanese Auto-mobile Producers," Chapter 21 In *Manufacturing Strategy*, John Ettlie, M. Burstein, and A. Feigenbaum, eds. (Boston: Kluwer Academic Publishers, 1990).

higher productivity, although the gap between Toyota and its competitors narrowed between 1973 and 1987.[11]

As another case in point, Northern Telecom found that as it simplified operations and shrunk manufacturing cycle times by more than 50 percent over a four-year period, its overhead costs during that same period dropped by 30 percent.[12] To see how this can happen, consider setup-reduction, the initial step in most JIT implementations. Reducing the time required to set up, or change over, a machine takes wasted time out of the production process; less machine downtime means more value added per hour, more output, and lower cost per unit. Per unit overhead costs also decrease because the costs are spread over more units of output. Northern Telecom's customer satisfaction index, a

---

[11]Ibid.
[12]Roy Merrills, *Harvard Business Review*.

EXHIBIT 2
1987 Productivity versus Throughput Time

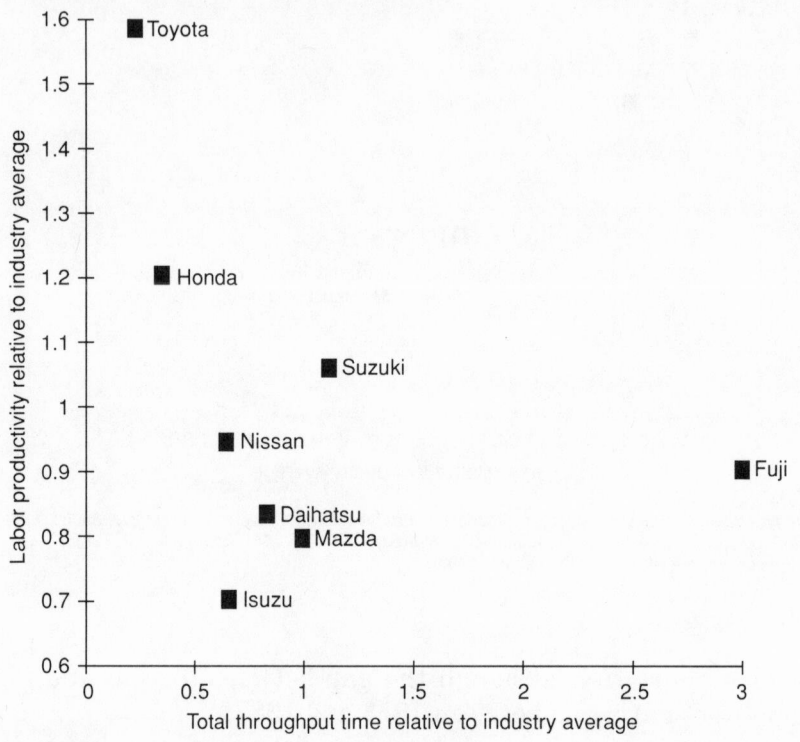

Source: Marvin Lieberman, "Inventory Reduction and Productivity Growth, A Study of Japanese Automobile Producers," Chapter 21 In *Manufacturing Strategy*, John Ettlie, M. Burstein, and A. Feigenbaum, eds. (Boston: Kluwer Academic Publishers, 1990).

proxy for quality, also increased over the same time period. Why quality improves as setup times decrease is discussed in Chapter 2.

A time-focused firm has as its goal *not* to waste time, but to use time as its most precious resource. Such firms seek to shrink the response time to customer's demand and, in effect, to make money by creating new demand for their products and services, new markets, and even new industries. These companies compress time and they create money. It is a conversion process. As a firm's response time improves, new markets are created and new customers are reached, which translate into money for the company.

Reshaping an Industry with Time

Photography illustrates how a business has been changed by time over the years. Sixty years ago consumers took snapshots with the Brownie, sent the camera itself and the film to Kodak, and waited patiently for their pictures. A good memory was required to recall the events captured on the prints received. In the 1970s, consumers took photographs with a conventional camera, waited until they finished a roll of film, took the film to the drugstore, and returned a week to 10 days later to retrieve the developed photographs of Johnny's birthday party. This gave real meaning to the phrase *time-lapse photography*. Into this time lapse stepped Polaroid with "instant photography" and, once again, instant gratification. The time lag inherent in conventional photography and its development process gave Polaroid a sizeable competitive advantage in the marketplace. And, for a while, the company exploited it.

In the past decade, time has been compressed in the field of conventional photography. Although this is not instant photography, the gap has narrowed significantly, as has Polaroid's time advantage. Photographs are developed in 50 minutes. Surprisingly, however, rolls of film with fewer frames have not been marketed. A shortened roll would diminish the long time lag that now occurs between the time that frame one is shot and the end of the roll. More frames encourage extra picture taking, however, and that probably explains why short rolls are absent from the marketplace.

Today's camcorders also provide instantaneous feedback: instant movies with sound and motion. Time-based competition clearly has revolutionized the way we record the past.

Other industries are ripe for transformation. The recorded music business is an example. Consumers no longer buy records; instead, they purchase music recorded on other media, including cassette tapes and compact discs. Digital tapes are just around the corner. The problem in this industry is the tremendous time lag between demand conception and consumption. Suppose a consumer has 15 favorite songs she wants to hear, all of which are recorded by different artists. Typically, she would have to go to the local sound shop outlet and purchase 15 different compact discs.

Here's a better way: The customer phones (or FAXes) the store with an order for her 15 favorite songs. The store enters the list of songs into a terminal and all 15 of them are coded onto a single compact disc. The customer picks up the disc at the store one hour later. The next step in the evolution of the process would be to use telecommunications to eliminate the visit to the shop. Digital transmissions of recordings to an in-home receiving device and electronic funds transfer would make the trip to the store unnecessary.

WHERE TIME-BASED COMPETITION WORKS

Service Industries: The Trailblazers

The trend toward a time-based strategy pervades service industries, which, by most measures, comprise the fastest-growing segment of the U.S. economy. Service industries, in fact, lead the manufacturing sector in the attention devoted to the time dimension. The explanation for this is simple. Providers of services have personal contact with their customers that, for the most part, manufacturers lack. Since services cannot be kept in inventory and must be delivered on demand, the service provider must be in intimate contact with the customer. Thus, in services, a consumer's preference for timely delivery is conveyed directly to the provider. In manufacturing, on the other hand, the manufacturer and the consumer often are separated by several layers of middlemen. Timely delivery is important to the product consumer, but the message is more difficult to transmit to the manufacturer, and often it is either lost in transmission or misinterpreted.

Fast-food firms are the prototypical time-focused competitors in the service economy. McDonald's charted the course for the industry and dramatically demonstrated that U.S. firms with a successful time-based strategy can dominate global markets, including Japan. Each element of the McDonald's operation is focused on time, and results show that by keeping wasted time out, costs can be kept at a minimum, and quality can be held at high levels. In the pizza segment, Domino's has used a time focus to become the number two chain in the United States. According to the CEO, Tom Monaghan, "Our whole business is built on speed."[13] And they deliver on that theme.

The time revolution also is transforming the financial services industry. Consider the last time you had a loan or a mortgage approved. How long did it take? A month? Six weeks? Would you not prefer the opportunity to patronize a time-focused financial institution, one in which the goal is instant customer gratification? To analyze the lending process, suppose that the lead time for a mortgage approval is four weeks. That equals 672 hours of the customer's time, spent waiting for a yes or no answer. During what portion of that 672 hours is work in the mortgage evaluation process actually being performed by a professional? Suppose instead that all the information on the mortgage application had been electronically encoded, a supposition that is not unreasonable. The amount of actual work involved probably would require no more than five minutes. To speculate about what happens to the mortgage application for most of that four-week period is needless because, whatever it is, no value is added to the service. All that is added is the uncertainty of waiting.

Times in financial services, however, are changing. Banks are taking a close look at their loan approval processes. The situation at Adca Bank in West Germany is described by the managing director, Hans Van der Velde: "Someone at the branch sent loan papers to someone at headquarters who was looking at them and changing them. Then his boss was looking and changing them. All those layers were useless."[14] Adca Bank subsequently cut the loan approval time from 24 to 12 manhours. Others have gone further: *USA Today* reported that Citicorp publicizes a 15-minute mortgage. This is as close to instant grat-

---

[13]Brian Dumaine, *Fortune*, November 6, 1989.
[14]Ibid.

ification as one can expect for a process that is notorious for its casual approach to customers' time.

The mail-order industry is another striking example. Time-based competitors emerged in this business and now clearly dominate the mail-order segment. How many times have customers completed an order blank for an item and read, to their dismay, "Please allow four to six weeks for delivery." The effrontery of that statement is incredible, even if the firm has a monopoly position in the market. Why should delivery take four to six weeks? Why should the customer allow it? Fortunately, those firms that take four to six weeks to deliver orders are subject to mail-order Darwinism. The industry's evolving time-based competitors—such as L. L. Bean and Lands' End—are driving predecessors into extinction.

The reason for this marketing success is the upheaval in the package delivery business, an upheaval fomented by time-focused competitors like Federal Express and UPS. These two firms compete head-to-head in the overnight delivery of parcels, and both firms have a time focus throughout their organizations—from the top deck to the loading dock. Later, in Chapter 10, CEO Fred Smith of Federal Express gives a behind-the-scenes look at how that firm was created and how it is driven by a time-based strategy.

In every segment of the service economy—even health care—time-based competitors are emerging and changing the nature of competition within their industries. Walk-in clinics and quick-response laboratories, both long overdue innovations in the time dimension of health care, are examples of increased emphasis on getting closer to the customer. The prescription eyeglass business has seen a remarkable time transformation. In the past, days were required to carry out ocular tests, grind lenses, and fit the frames. Now customers can get the entire package in under an hour—the same amount of time required to have photographs developed.

Time Compression in Manufacturing

As in service industries, a time-based competitor is emerging in virtually every manufacturing segment. With its focus on intimate customer interaction, however, the changes in services' response time have been more immediate and obvious. Time-based practices are spreading in manufacturing, but they may not be as evident because a firm's quickness in manufacturing can be partially obscured by dilatory distribution systems and sluggish customer service. So the time-based changes cannot be seen as dramatically as in service businesses. In addition, most of the pacesetters in manufacturing tend to be Japanese, and U.S. firms, though catching up, have responded slowly. But this is changing.

General Electric is an example of how an industrial giant can become a more timely competitor. Faced with stiff competition in circuit breakers from firms such as Siemens, GE recognized that it needed faster response to customer orders. To gain speed, the company consolidated the production of circuit breakers into one plant in North Carolina. In addition, GE greatly simplified the plant's products: over 26,000 of its 28,000 components were eliminated, while leaving customers with choices among 40,000 product variants.[15] Flexible, automated manufacturing cells were installed to carry out production.

The results of GE's restructured manufacturing, as reported in *The Wall Street Journal*, are impressive and instructive. Productivity increased by 20 percent and costs decreased by 30 percent in the same year. More important from the customers' point of view, orders required only three days to fill instead of the former three weeks. The plant's ROI exceeded 20 percent, and GE gained market share in a moribund market. According to William Sheeran, general manager for GE, "We had to speed up or die."[16]

The Allen-Bradley Company demonstrates how an old-line manufacturer from the Rust Belt can be transformed and revitalized through a time-compression strategy. This 80-year-old Milwaukee-based business initiated a major effort to capture a larger share of the industrial controls and industrial automation market. In the early 1980s, Allen-Bradley's market dominance in electric contactors was threatened because of foreign competi-

---

[15]Timothy D. Schellhardt and Carol Hymowitz, "U.S. Manufacturers Gird for Competition," *The Wall Street Journal*, May 2, 1989.
[16]Brian Dumaine, *Fortune*, November 6, 1989.

tion, coupled with emerging international standards for motor controls. Led by its president, Tracy O'Rourke, the firm redesigned its contactor facility in Milwaukee as part of a strategic plan to capture a higher share of the world market for electrical contactors. The plan called for a high-volume flexible facility that could receive, manufacture, and ship orders within 24 hours—that is, make to order within 24 hours. Essentially, Allen-Bradley aspired to become the Federal Express of the contactor market.

What happened? The company achieved its time compression targets, customers responded with purchases, and Allen-Bradley attained the lead position in the global market. The relative cost per unit dropped to 60 percent of that required in previous conventional manufacturing. The cost of production, in fact, became the lowest in the world. Instead of the 125 product variations made in the former plant, the facility now makes over 600 product variations to order. The return on assets reached five times the level possible with traditional manufacturing. And quality did not suffer: The reject rate runs only 20 units per million. The firm's resurgence in the basic electric contactor and relay market proves what a time-focused competitor can achieve.

In automobiles, Honda has capitalized on its industry-leading speed in introducing new products. A chilling comparison is provided by two newspaper articles that appeared on successive days in November 1989. A November 25 article in the *New York Times* described how Japanese auto firms get a boost in sales by offering more advanced engines than their U.S. counterparts.[17] The conventional engine from Detroit has two valves per cylinder, whereas most of the engines in today's Japanese cars have four or more valves per cylinder. More valves mean more powerful, more fuel-efficient engines. According to the *New York Times*, Christopher Cedergren of the auto-industry research firm of J. D. Power & Associates contends that "it's important for domestic manufacturers to get these engines into vehicles aimed at younger buyers. The domestics are stalling for time."[18]

---

[17]Doron P. Levin, "New Japan Car War Weapon: A 'Little Engine That Could," *New York Times*, November 25, 1989.
[18]Ibid.

Many of the engines produced by the so-called Big Three automakers in Detroit were designed during the 1960s. These manufacturers steadfastly claim that the multivalve technology is
"inappropriate for the American market" and that many U.S.
consumers prefer the conventional engines.[19] The continually
eroding market shares of the Big Three cast doubt on that claim.

An article in *The Wall Street Journal* on the preceding day
told the rest of the story—that is, why Detroit automakers cannot get the multivalve engines developed in time. This article
described the "cup holder war" currently being waged in Detroit
among producers of the new minivans. GM's new APV minivan
with 14 built-in cup holders is the clear leader in the competition. Ford apparently plans to rally. The *Journal* account quoted
a director of interior design at Ford as declaring, "You can bet
cup holders are a strong feature we will be considering for all
vehicle segments."[20]

SPEED: THE COMPETITIVE ADVANTAGE

Speed kills, but in business speed is fatal to the other guy: the
competition. As P&G demonstrated in consumer products, quick
response to market demand is one feature that most market
leaders have in common. Customers want the newest products
and services, and the companies best positioned to satisfy those
needs make the most sales.

Although the motivation for a time-based corporate strategy is provided by marketing's need to get closer to customers,
the key to becoming a time-based competitor lies within the organization. From the customer's point of view, what matters is
the *total* time required to deliver the product or service. To compress time, fundamental changes must be made in every function that affects the delivery of the product (or service) to the
customer.

To their credit, U.S. firms have worked assiduously to reduce the time required in their manufacturing processes. By

---

[19]Ibid.

[20]Melinda Guiles and Neal Templin, "Detroit Gets Serious about Assuaging America's
Thirst," *The Wall Street Journal*, November 24, 1989.

most measures, the Just-In-Time (JIT) revolution of the 1980s has been successful. Many leading U.S. firms achieved short cycle times in manufacturing and reduced inventories; in effect, they became "lean and mean" manufacturing firms. For example, Motorola can now build to order and ship an electronic pager within two hours; several years ago the task consumed three weeks.[21] GE, similarly, has taken days out of the production cycle for dishwashers and washing machines.

Manufacturing, however, is but one link in the value-delivery chain. To the customer, time saved in manufacturing is no more and no less important than time saved in distribution or customer service. Customers want the product as soon as possible after the need emerges: instant gratification. If the time saved by short-cycle manufacturing is later lost in the delivery process, the customer is unimpressed—especially if the product is outmoded upon delivery.

Many firms, through time compression of parts of their operations, have successfully shortened production cycle times and enforced frequent deliveries from nearby suppliers. But many of these same firms, on the other hand, still have snail-like customer service functions. A product may be transformed from raw material to finished product in a matter of hours, but days or even weeks may be needed to get the order from the factory floor to the customer. Their products, after being manufactured rapidly, take too long to navigate the labyrinth of the distribution system.

Most critical in competitive markets is the speed with which the firm brings new products to market. In the highly competitive auto industry, Japanese firms, such as Honda, sustain an advantage by virtue of the rate at which they can introduce new technology. These firms not only are faster, they are also more efficient. Kim Clark of the Harvard Business School reported that during the 1980s Japanese automakers, on average, turned out an automobile development project in 50 to 60 percent of the engineering hours required by the Big Three automakers.[22] Ac-

---

[21]Brian Dumaine, *Fortune*, November 6, 1989.

[22]Kim B. Clark, "Project Scope and Project Performance: The Effect of Parts Strategy and Supplier Involvement on Product Development," *Management Science* 35, no. 10, pp. 1247–63.

cording to Vladimir Pucik of the University of Michigan, "The game the Japanese are going to play is to leave the Americans building well-engineered but boring and obsolete cars. It's the next battlefield."[23] Some U.S. firms in other industries have launched time-focused programs to counter with rapid new-product introduction cycles of their own. In Chapter 6, John Bailey describes how Honeywell Building Products uses team concepts to shrink the new product development cycle. Mohan Kharbanda explains Xerox's approach to new product development in Chapter 7, and Stephen Hamilton describes Hewlett-Packard's efforts in Chapter 8.

Many U.S. firms have compressed time in one function, but few have successfully eliminated wasted time in all functions. Few firms focus on the total response time and get all of the pieces right. The evidence strongly suggests that, across most organizations, wide-open windows of opportunity for removing wasted time can be found in every function. According to George Stalk of the Boston Consulting Group, in the average firm less than 5 percent of the total time required to manufacture and deliver a product to a customer is spent on actual work.[24] Research carried out by graduate students at the Owen Graduate School of Management at Vanderbilt University supports Stalk's observations.[25] In one case a boot manufacturer required seven weeks to convert an order for a pair of boots into the finished product, but less than 3 percent of that time was devoted to actual value-added work on the boots. In another case, a telecommunications firm needed six to eight weeks to turn a customer's order into a specific work order ready for the factory floor.[26] By bird-dogging the paperwork for a large sample of these orders, the student-researchers found that idle time constituted more than 95 percent of the order-processing interval.

The question is: How to do it? How does a firm achieve a

---

[23]John Bussey and Douglas Sease, "Manufactures Strive to Slice Time Needed to Develop Products," *The Wall Street Journal*, February 23, 1988.

[24]George Stalk, "Rules of Response," *Perspective: Time-Based Competition Series*, Boston Consulting Group, Boston, Mass.

[25]Randy Mead and Joseph Blackburn, "Time-Compression at Acme Boot Company," Owen Graduate School of Management Working Paper, Vanderbilt University, 1987.

[26]Susan Peppers, "Time-Compression at Northern Telecom," Research Report, 1988.

pervasive time focus, and can the same time-compression techniques be used in different functions? Fortunately, a set of guiding principles that apply across the organization exists for the time-based competitor. Principles that shrink time in manufacturing also can be applied successfully to hasten new-product development, distribution, customer service, and so forth. Firms that devise successful short-cycle manufacturing can achieve responsive customer service, rapid distribution, and quickened product development. Moreover, the guidelines apply equally well to paper-flow processes in offices and to service delivery processes in service industries. The key to understanding ways to remove time can be found by tracing the origins of time-based competition.

This book serves two prime purposes. First, it describes the transformation that must occur for a firm to become a time-based competitor. The techniques needed can be learned; the technology is transferable. A firm's motivation to focus on speed, however, must come from the top of the organization and, in many firms, this requires a change in corporate philosophy or corporate culture. To be succesful, firms must adopt time as a strategic paradigm and that requires a management commitment. Subsequent chapters will examine how the management of leading firms—such as Hewlett-Packard, Milliken and Company, Honeywell, Federal Express, and others—applied time as a competitive strategy.

Second, the book discloses the techniques of time compression that can be applied to each step in the value-delivery chain and, if applied, reduce the total time required to serve the customer.

ORIGINS OF TIME-BASED COMPETITION

The phrase *time-based competition* (TBC) originated with George Stalk and his colleagues at the Boston Consulting Group. In researching their book, *Kaisha: The Japanese Corporation*, James Abegglen and George Stalk observed first-hand the evolution of JIT production systems at companies such as

Toyota.[27] The authors noted how these JIT firms learned to time-compress manufacturing and then propagated skills in time reduction to the rest of the organization. Firms that achieved speed in manufacturing soon added a responsive distribution network, rapid new-product development cycles, timely customer service functions, and so forth. The JIT innovators became the world's first time-based competitors.[28]

Stalk and Abegglen deduced from their observations that TBC is the extension of JIT principles into every facet of the value-delivery cycle, from research and development through marketing and distribution. They noted, further, that when one firm achieves a significant time advantage in product or service delivery, the nature of competition within that entire industry is changed. Cost becomes secondary to response time. This impact occurred, for example, in the textile and apparel industry, where TBC capability has been dubbed "quick response."

Both JIT and TBC have identical objectives: Eliminate all time waste in the production of a product or delivery of a service. Just-in-Time is characterized by small production runs, quick changeovers, and low inventories—all of which escalate product velocity. Applying these principles properly maximizes the percentage of time in which value is added to the product. These qualities also characterize TBC: Eliminate idle or dead time wherever it exists, process work in small batches, and maximize the value-added time. However, TBC goes one step further than JIT and encompasses not only manufacturing but also the complete value-delivery chain of the product or service.

Why is the focus on JIT important? A popular misconception is that JIT consists of reduced inventory and frequent shipments from suppliers. It is much more than that. The real secret to a successful JIT firm is an emphasis on time—eliminating time . . . eliminating waste . . . moving every phase closer to the customer in time . . . in distance . . . moving closer to their tastes and reacting more quickly to changes in taste.

---

[27]James Abegglen and George Stalk, *Kaisha, The Japanese Corporation* (New York: Basic Books, 1985).

[28]In Chapter 3, George Stalk provides updated insights on the evolution of time-based competition.

The ongoing JIT revolution is perhaps the first manifestation of a new type of time-based competition. JIT, however, is not an end in itself for a firm; rather, it is an evolutionary step toward the long-term goal of total time compression. Zero inventory and proximity to suppliers, while important, have only limited impact on profits. The real benefits—the ones that provide a sustainable competitive advantage—result from shrinking the duration of the whole manufacturing cycle. Cycle-time compression translates into faster asset turnover, increased output and flexibility, and satisfied customers. Thus, the diminished inventory often associated with JIT is more of a side benefit than a driving force.

EMPHASIZING HOW TO DO IT

The objective that guided this work was to create a "how-to" book for the would-be time-focused competitor. To this end, this book examines how leading U.S. firms have used time reduction as the centerpiece of their competitive strategy. Five chapters depict a particular firm's struggle to become a time-based competitor and how this effort enhanced the company's ability to compete effectively both at home and abroad.

Virtually all of this material, moreover, derived firsthand from managers of these corporations, who presented accounts of their firms' experiences at a conference on time-based competition held at the Owen Graduate School of Management, Vanderbilt University, in December 1988. These insider accounts provided fresh raw material for learning: for example, how Federal Express became a leader in the overnight package delivery business; how Hewlett-Packard eliminated excess time from its new-product introduction processes; how Milliken led the "quick response" crusade in the textile industry; and how Xerox and Honeywell attacked the problem of long lead times in product development. The undergirding thesis of this book is that TBC is the logical extension of the JIT principles that have revolutionized manufacturing to the other parts of the product delivery chain: new-product development, engineering, customer service, distribution, and the rest. Moreover, from the customer's viewpoint, these parts are equally important because any day saved in delivering a product is equally valuable. It does not matter to the customer how or where time was compressed. It matters only that the product is available when wanted.

Becoming a time-based competitor is difficult for many firms because a focus on time is unnatural. The focus has been on cost and, lately, on quality. All the performance measures— from the traditional cost-accounting system to newer quality goals—lack an emphasis on speed. So what is required is nothing less than a phase shift, a quantum shift within the corporation to new rules, new goals, and new performance measures. For the firm that completes the mission, however, the results justify the effort. Experience shows that a time-focused firm creates an upheaval in an industry. It stands the competition on its head and dominates markets.

CHAPTER 2

JUST-IN-TIME: THE GENESIS
OF TIME COMPRESSION

Joseph D. Blackburn

Although Just-In-Time (JIT) has been a potent force in revolutionizing worldwide manufacturing, only a fraction of its potential has been realized. Outside the factory walls, JIT is in its infancy. Significantly greater benefits will accrue to those firms that apply the JIT philosophy to other parts of the value-delivery chain—new-product development, engineering, customer service, logistics, and distribution. "White-collar JIT"—the application of JIT to paper flow and information processes in offices—is emerging as a new challenge for management.[1]

Using JIT as a tool merely to shorten manufacturing cycles is much like running an eight-cylinder car on three cylinders: The full power of the concepts is underutilized. This chapter asserts that JIT concepts are the key to the entire time-compression process. This and subsequent chapters show that the same steps taken to implement JIT on the shop floor can be used to time-compress other segments of the delivery chain.

Unfortunately, the transfer of JIT concepts beyond the factory walls is blocked by narrow and conventional notions of JIT

---

[1]Julie A. Heard, "JIT for White Collar Work—The Rest of the Story," Chapter 12 in *Strategic Manufacturing; Dynamic New Directions for the 1990s*, Patricia E. Moody, ed. (Homewood, IL: Dow-Jones Irwin, 1990).

applications. Many managers view JIT through blinders inadvertently provided by some of the so-called JIT experts who stress the virtues of JIT only as a system for inventory reduction and material control. Consequently, important functions such as new-product development and customer service, which have neither inventory nor materials, appear to many managers as not being fruitful areas for applications of JIT principles. In addition, this perspective overlooks the major benefits that JIT applications can produce: time compression, space reduction, quality improvement, and operations with the flexibility to deliver a wider range of products or services. In functions other than manufacturing, moreover, these benefits are the only ones that accrue from JIT. These benefits enhance customer satisfaction and that, in the long run, produces a more attractive profit-and-loss statement.

For any firm to realize the complete package of JIT benefits, its management focus must be on time, not inventory. This different—perhaps radical—outlook enables managers to grasp the full potential of JIT within an organization and to transfer the technology beyond manufacturing. The writer contends that the forces behind JIT are not driven by manufacturing. On the contrary, JIT was conceived and evolved in Japan, primarily at Toyota, as a reaction to market forces—specifically, the consumer's dual needs for product variety and quick response. Satisfying these needs demands a manufacturing process that is both time-compressed and flexible.

Just-in-Time provides manufacturing with the flexibility and speed essential to meet global competition. By examining the steps required to achieve JIT, managers obtain insights that enable them to develop similar capabilities in nonmanufacturing functions, including new-product development.

HOW JUST-IN-TIME CAME TO AMERICA

Conventional U.S. manufacturers in the early 1980s were eager to try anything new that promised an escape from their moribund state. The traditional batch processor or manager of a job shop was saddled with excess work-in-process (WIP) inventory, shoddy quality, long lead times, and irate customers. The expensive material requirements planning (MRP) system installed to solve these problems was more like Pandora's box than a panacea. Domestic competition was fierce, but the prospect of offshore competition was frightening.

Into this environment JIT arrived with a flourish from the Orient and was greeted by a manufacturing community grasping at straws and seeking anything to enhance its competitive position. With the sole exception of the techniques of total quality control (TQC), no concept in the 1980s captured the hearts and minds of manufacturing managers as did JIT. Within manufacturing, JIT assumed the trappings of a religion. Pilgrimages in the form of plant tours were made to Japan and other locations where, reportedly, "manufacturing miracles" had transpired. Based on their visions, industrial gurus and evangelists traveled the countryside to spread the word to the uninitiated. Seminars and gatherings were held in exotic locations to provide deeper understanding. The word also spread through books, articles, speeches, video tapes, and even audio tapes that enabled managers to learn about JIT while driving to and from work.

Cynics, however, noted that today's JIT experts are yesterdays's MRP experts. The same consultants who touted MRP in the 1970s and early 1980s became the chameleons of the manufacturing consulting world. Overnight, they altered their philosophies and principles to blend with the latest rage on the manufacturing scene.

U.S. firms' overemphasis on JIT's inventory reduction and material control aspects is attributable to the MRP-based myopia of these early experts. As experts in material control and shop scheduling, they naturally viewed JIT through the same lenses and missed the wider benefits. They looked at Toyota's JIT system and saw only Kanban cards and inventory reduction. These consultants then transferred their narrow vision to eager, receptive audiences in the United States. In those frenetic years, moreover, many of the consultants' messages about JIT were garbled in the transmission or misinterpreted by the recipients.

As a result of this incomplete translation, many managers today have a narrow view of JIT. A survey by Coopers and Lybrand indicated that many manufacturers considered the important features of JIT to be reduced inventory and frequent deliveries from suppliers.[2] Another survey reports that "JIT can be viewed as a system that links a supplier to a buyer/customer with responsive delivery."[3] Reduced inventory is an important JIT benefit and a critical visual measure of progress, but it is not the overriding objective.

Manufacturing executives were not hoodwinked by the JIT gurus. They were willing coconspirators; they desperately sought a "quick-and-dirty" solution to their problems and seized upon the superficial aspects of JIT. Manufacturers with inefficient and obsolete plants, looking for a bandaid or scapegoat, focused on extracting frequent deliveries of raw material from suppliers as the key JIT concept. This provided a convenient way to shift the blame to suppliers and away from the dinosaurs in their own shops. Leaning on suppliers to hold the inventory and speed up delivery is easier than making substantive operational changes at the source of problems within the factory walls.

The purpose here is not to cast aspersions on either the JIT messengers or their audience, but to stress that a focus on inventory reduction misses several points. Reduced inventory is a relatively small benefit compared to the other benefits achievable through JIT. Moving suppliers closer to the factory usually results in little more than shifting inventory costs further back in the distribution chain. This is the "Revlon approach"—a cosmetic change which alters nothing that contributes to the ultimate cost of production. Without an accompanying time reduction in the manufacturing process, moreover, slashing inventories can be catastrophic. Consider the fate of a company operating with an inefficient and unresponsive manufacturing process coupled with insufficient stocks of inventory to meet customer demands. This "double whammy" causes customer service

---

[2]Robert W. Hall, "Just-In-Time Manufacturing: Discovering the Real Thing," *Corporate Strategies*, February 1987, pp. 3–6, Tocqueville Asset Management Corporation, New York.

[3]Albert F. Celley, William H. Clegg, Arthur W. Smith, and Mark A. Vonderembse, "Implementation of JIT in the United States," *Journal of Purchasing and Materials Management*, Winter 1986, pp. 9–15.

to suffer more today than in pre-JIT days. As customer complaints mount, managerial support for JIT withers.

U.S. firms' restricted focus on inventory control and delivery issues yielded other unwelcome side effects. First, many managers dismissed the initial reports of JIT as consultants' hype and were slow to react because—having been burned earlier by their failure to implement MRP systems successfully—they were chary of JIT benefits promised by the same consultants who led their ill-fated MRP journey. Second, managers who acted upon the consultants' advice and implemented JIT focused their efforts and performance measures in the wrong places: on inventory reduction. When performance failed to live up to expectations, JIT became the baby that is thrown out with the bath water.

JUST-IN-TIME: ACCENT ON TIME

Quick response to customer demand is one of the major benefits of JIT. Time, or speed, is the linchpin of this manufacturing philosophy. Inventory, on the other hand, is an ancillary benefit. In fact, time is the key to comprehending the wider application of JIT principles. As a time-compression process, the principles of JIT become portable: They generalize to other parts of the organization and its value-delivery chain. The JIT principles can be applied to the new-product introduction process, to customer service, to logistics and distribution, and even to service industries in which inventories are nonexistent.

Much of the writing on JIT is narrowly focused and misses the overarching objective. A strategic overview is lacking. What is accomplished with JIT? By examining its evolution, this chapter illustrates how JIT arose in Japan out of a need for better coordination between manufacturing and marketing. To be customer-focused, manufacturing and marketing must have synchronized objectives. To the customer and to marketing, this means quick response time and a wider variety of product offerings. To manufacturing, it means speed and flexibility. Subsequent sections reveal that JIT evolved in response to these dual environmental requirements.

VARIETY IN THE PRODUCT LINE: A CATALYST FOR JIT AND TIME COMPRESSION

Today's markets are global. A vignette in Chapter 1 described how consumer giant Procter and Gamble redoubled its efforts to compete in foreign markets in order to maintain a dominant position in most of its product categories. To preserve historical growth patterns, U.S. manufacturers must sell their products globally. As the U.S. market is carved up by an increasing number of foreign players, domestic manufacturers can no longer sit back and reap their accustomed share. It is axiomatic that increased competition and new contestants within a market will reduce the shares and profits of the established rivals. In the 1980s, the Big Three automakers witnessed their share of the U.S. auto market drop from a peak of 72 percent in 1985 to approximately 67 percent in 1989.[4] Likewise, U.S. producers of machine tools, who owned the market in the early 1970s, have seen Japanese toolmakers, such as Yamazaki, consume growing chunks of the U.S. market. From 1976 to 1981, imports of Japanese machine tools to the United States grew tenfold from $67 million to $687 million. In 1987, over 50 percent of machine tools purchased in the United States were imported, and the Japanese supplied over half of these imports.[5]

The "catch-22" is that global markets are highly heterogeneous; variety is the salient feature. Although there are regional differences in product preferences between New England and New Mexico, these differences are insignificant compared to consumer preferences in Paris, France, and Paris, Tennessee. To their dismay, U.S. manufacturers realize that they cannot merely substitute new language on the package and send it overseas. They can—but it won't sell.

Because tastes and standards are different, a product must

---

[4]Data from Sanford C. Bernstein; reported by Joseph B. White in *The Wall Street Journal*, January 5, 1990, p. B1.

[5]Anthony J. Zahorik, Timothy L. Keiningham, and Joseph D. Blackburn, "Time Competition in the Machine Tool Industry," Owen Graduate School of Management Working Paper, December 1989.

be produced in different forms for different corners of the global market. Automobiles may require different emission devices or different locations of the steering wheel. Despite efforts to agree on standards, electrical systems also differ around the world. Even within national borders there is scant agreement on data standards in computers and telecommunications. While most of the world has accepted a metric standard, some have suggested that the United States is merely "inching along" in that direction.

To compete in foreign markets, the U.S. manufacturer must produce a more complex product line, which translates into greater variety among its product offerings. Packaging and labeling alone are not enough. Expanded product lines have become imperative for the corporate strategy of firms engaged in worldwide competition. And therein lies the problem for manufacturing. It must quickly develop the expertise to produce a wider range of products to be congruent with the goals of marketing and the rest of the organization.

Variety and Complexity: A Marketing-Manufacturing Conflict

In his *Harvard Business Review* article, Benson Shapiro asked the provocative question "Can Marketing and Manufacturing Coexist?"[6] The article emanated from Shapiro's observation that, in most industrial organizations, the manufacturing and marketing divisions do not always act in harmony and, in fact, engage in frequent minor skirmishes and occasional pitched battles. The reason for this long-festering conflict is obvious: Marketing and manufacturing have opposing objectives (see Exhibit 1). Marketing typically is rewarded based on sales and, consequently, sets its sights on satisfying customer demand. Manufacturing usually is measured on cost performance and, therefore, zealously pursues that objective through its production systems.

---

[6]Benson P. Shapiro, "Can Marketing and Manufacturing Coexist?" *Harvard Business Review*, September–October 1977, pp. 105–14.

EXHIBIT 1
Market and Manufacturing: Conflict at the Interface

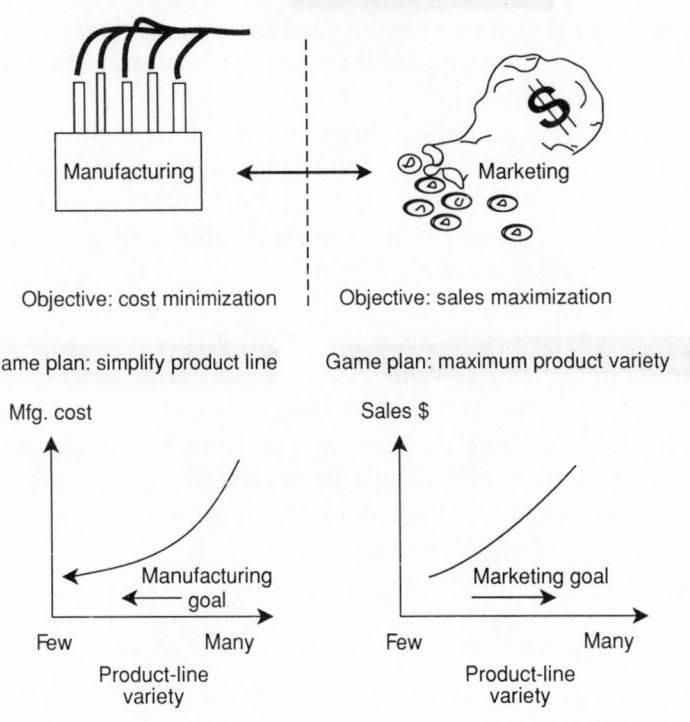

This internal conflict poses a major dilemma for a firm. Marketing seeks maximum variety in the product line in order to present the widest possible appeal to consumers; the marketer's dream is a plant that could turn out the following sequence: a microwave oven, a dishwasher, a tractor, another microwave oven, and then a 757 jetliner. At the other extreme, manufacturing's ideal is to churn out the same commodity product in a single color. A company's dilemma is, thus, variety or volume: scope or scale.

The dilemma is not easily resolved for a conventional manufacturing facility. Product-breadth decisions become a tug-of-war game between divisions, with top management frequently called in to referee. Long lead times and fixed production schedules are the norm, and the typical plant lacks the flexibility to produce competitively the variety that marketing desires.

A diesel engine producer with four types of diesel engines—models A, B, C, and D—provides a good example of how interdivisional conflicts can arise. Exhibit 2 shows how manufacturing produces the four models.

The firm produces in large batches because long changeover times are required to shift production from one engine type to another. To increase efficiencies when the machines are set up for a particular model, long production runs are required. Large batches mean larger piles of WIP and longer lead times in the shop.

Typically, marketing lands a larger order for model C engines from an important customer who needs immediate shipment. A check with the warehouse indicates that not enough units are available to meet the order. The sales manager asks manufacturing when model C will be produced again, only to learn that it is not scheduled for another two weeks. The rest of the conversation goes something like this.

MARKETING:

"Well, what are you guys producing now?"

EXHIBIT 2
Production Schedule for Diesel Engines

Week		1					2					3					4		
Day	1	2	3	4	5	1	2	3	4	5	1	2	3	4	5	1	2	3	

Product

A — Set-up (3/4 day) — Production (week 2, days 2–5)

B — Set-up (1/2 day) — Production (week 1, days 2–5)

C — Set-up (1/3 day) — Production (week 3, days 2–5)

D — Set-up (1/2 day) — Production (week 4, days 1–3)

MANUFACTURING:

> "Model B—the rest of the week. You know our schedule; we send it to you every Monday based on your forecasts."

MARKETING:

> "But we've got too much of model B in stock already. Can't you guys produce what we need? This is our most important customer."

MANUFACTURING:

> "Why can't you guys in sales learn to forecast? It takes an entire shift to set up for another model and we just can't afford the lost production this near the end of the month. We can't change the production schedule every time one of you guys sneezes."

MARKETING:

> "The next sound you hear from us won't be a sneeze!"

This story has many different endings, and almost all of them are unpleasant and tend to destroy an important relationship between two functions in the organization that need to dance together. Escalating conflicts between marketing and manufacturing turn the interface between the two functions into a demilitarized zone and prohibit the execution of a coherent corporate strategy.

The Focused Factory

What are the practical consequences of expanding the product line? While marketing loves it and manufacturing hates it, variety in the product line means a more complex operation on the shop floor: more changeovers, more parts to track, and more complex schedules. Manufacturing managers despise complexity because it is more costly to produce and more difficult to manage. Every manufacturer desires a product line like Henry Ford's Model T line: every unit was identical—"any color as long as it was black." Manufacturers seek to focus on production numbers. Their performance measures improve when they gain economies of scale, which comes from long production runs of a single product. On the other hand, product-line variety tends to create diseconomies of scale on the factory floor: short production runs, frequent changeovers, and complex schedules.

Faced with a strategic choice between diversity and focus, manufacturing executives found a slogan to rally behind in the "focused factory"—a concept made famous by Wickham Skinner in his 1974 *Harvard Business Review* article.[7] The "focused factory" dominated manufacturing strategy in the late 1970s and has many proponents even today. Skinner extolled the virtues of a focused manufacturing operation: simplify the product line and wield an efficient, simplified production operation as a competitive weapon. Manufacturing managers embraced the concept because it made their jobs easier and they spearheaded campaigns to build the entire corporate strategy around focus.

Among the firms that embraced the strategy was Chrysler Corporation. In the mid-1980s, CEO Lee Iacocca adopted a focus strategy and made drastic cuts in the number of car platforms and option combinations available, effectively reducing the number of variations available to the customer.[8] For Chrysler, the effect of the strategy was short-lived: Focusing down the product line gained short-term cost reductions, but Chrysler was unable to sustain the advantage in the face of increased product variety by automotive competitors. For firms in other industries that did not choose a corporate strategy based on focused production, the level of internal conflict merely escalated because marketing remained vigorous in its pursuit of a strategy to offer greater variety to consumers. Focused production and greater product variety are collision-course strategies. The only ways to focus production and gain variety are either to add dedicated production lines or develop extremely clever product designs.

Fast-food merchandisers are caught in this classic focus-variety squeeze. McDonald's, Burger King, and Wendy's are under constant market pressure to increase the variety of their menus. Twenty years ago McDonald's offered any sandwich you wanted

---

[7]Wickham Skinner, "The Focused Factory," *Harvard Business Review*, May–June, 1974.

[8]Liz Roman Gallese, "Counselor to the King," *The New York Times Magazine: The Business World*, September 24, 1989.

as long as it involved a hamburger. Customers, however, demanded more and satisfied those demands at competing fast-food outlets. McDonald's responded by adding fish sandwiches, special hamburger offerings, salads, and even breakfast items. Today, it offers rib sandwiches and pizza in selected test-market areas.

Competition in the fast-food business forced firms to adapt to a variety strategy. Some competitors with focused production operations, such as Burger King, struggled with the addition of variety, particularly in the breakfast market. Every fast-food purveyor has moved away from, if not abandoned, a strategy based on focused production, and this created a more complex food-preparation operation.

Coping with Complexity through Flexible Manufacturing

Manufacturing vice presidents in the United States staunchly defend focused production. They resist efforts to add products to the line because these products drive up production costs and take foods off their tables in the form of performance bonuses. Unless aggressive actions are taken, complexity tends to drive up the breakeven volume.

Two automobile manufacturers located near Nashville, Tennessee, deal in different ways with the cost of complexity: the General Motors Corvette plant in Bowling Green, Kentucky, and Nissan Motor Manufacturing USA, a so-called Japanese "transplant," in Smyrna, Tennessee. Both plants produce cars and—because consumers disagree about color—both plants must paint them different colors. Painting is an expensive process. The changeover from one color to another is difficult, time-consuming, and costly. Car makers would prefer to avoid changeovers and produce all cars in a single color; it does not even have to be black.

The Corvette plant (which is the older of the two plants) paints cars in the conventional way. Costly changeovers are minimized by producing in large batches. For a day or two, the plant produces baby-blue Corvettes, then fiery-red ones, a short run of yellows, a batch of whites, then blacks, and then the sequence repeats. The most popular colors are produced in the largest batches. Basically, however, this means that a particular color is produced no more than once a week. On a given day, the visitor would see only a single-color Corvette, perhaps two colors, on the assembly line. Eighty miles away at the Nissan plant, the picture is strikingly different. The colors of the cars (and trucks) exiting a single assembly line follow no particular pattern. The color sequence appears to be random—a blue one, followed by a red, then a black, and so on. Painting here can be in batches of one!

This comparison is unfair because the vehicles are not of comparable value nor are the plants of equal age. Also, painting a Corvette may be a more exacting process than painting a Nissan Sentra, an automobile whose cost is less than a third of the Corvette's. It is clear, however, that Nissan attacked the cost of variety and achieved a solution that allows for mixed-color assembly without large-batch production. In fact, the apparently random sequence of trucks and cars on the line indicates that Nissan has achieved small-batch, mixed-model assembly as well. Batch sizes of one give Nissan the flexibility to respond quickly to changing conditions. Manufacturing's capability supports the company's market-driven strategy.

The cost of complexity (variety) is the cost of changing over from making one product to making another. These are the diseconomies of scope. Dedicated lines producing a single product require no changeover costs, not unlike Ford's Model T. In conventional manufacturing, changeover means shutting down machines to reset them for the new product: dies must be changed, new tools are needed in the workplace, new colors must be added, jobs must be changed, new machine settings are required, and so on. Idle time results, and the cost of the product is driven upward.

Evolution of Flexible Manufacturing

Variety and complexity create a strategic dilemma because they are opposite sides of the same coin. Variety is a boon to marketing and the bane of manufacturing. Marketing regards variety as a benefit because variety is what the customer demands, and variety becomes one of marketing's strategic objectives. Manufacturing uses the term complexity to describe variety because, on the factory floor, variety in the product line begets complexity. Producing a large number of products in the same facility dictates complex, fluctuating schedules; changeovers of machines from one product to another; adaptive skills in the work force; and so forth.

The Nissan example, however, reveals that firms can cut the Gordian knot posed by complexity. Instead of relying on focused, rigid production systems, firms intent on eliminating the cost of complexity—the costs of a changeover—must develop increasingly flexible systems. Nissan achieved flexibility in painting by eliminating much of the time in a changeover from one color to another. In painting, sophisticated technology is required for rapid changeovers: electrostatic paints; quick-flush systems; and powerful, effective ventilation procedures.

Reducing the changeover time is the initial step on the road to a flexible manufacturing system that is the result of a successful JIT implementation. Technological "fixes," such as Nissan employed, are frequently unnecessary to reduce setup times. Later in this chapter, in our explanation of the origin of JIT, we will explain how most changeovers can be time-compressed using single techniques first developed in Japan for die changes.

During the late 1970s and early 1980s, manufacturers followed two divergent paths on the issue of focused versus flexible production. While the guiding slogan for many U.S. firms during that time was "focus your factory," firms in other parts of the world attacked the flexibility problem head-on and with startling results. James Abegglen's and George Stalk's studies of Japanese manufacturing made it clearly evident that advances in manufacturing processes had been made by the leading Japanese firms, including Toyota, Sony, Mitsubishi, and others.[9] Sig-

---

[9]James C. Abegglen and George Stalk, Jr., *Kaisha, The Japanese Corporation* (New York: Basic Books, 1985).

EXHIBIT 3
Production Assembly Sequence, Yanmar Diesel (1976)

Date	A	B	C	D	Total
		Product Line			
2/1	320				320
2/2	320				
2/3	320				
2/4	120	200			
2/5		320			
2/6		320			
2/7			320		
2/8			320		
2/9				320	
				320	320
Average Days of Inventory	3.2	2.9	3.5	3.5	

Source: The Boston Consulting Group; Yanmar Diesel Engines.

nificantly, however, the advances in Japan in the 1970s were not restricted to a few major players but, instead, were quite widespread.

Consider, for example, the data from a lesser-known Japanese firm, Yanmar Diesel. Exhibit 3 is a 1976 snapshot of a shop production schedule for Yanmar Diesel. This schedule is not unlike what one would find on the desk of an American production manager during the same time period. Yanmar was involved in conventional batch manufacturing in which several shifts were devoted to the production of a single model because of the cost of setting up for the next model. The changeover cost drove Yanmar to long production runs.

Another snapshot of Yanmar's production schedule, taken six years later, is provided in Exhibit 4. It is an astounding contrast to the 1976 snapshot. The batch size has been reduced to one, and products could be produced in any sequence. Yanmar developed a production system—in the space of six years—that made possible batch sizes of one. Work-in-process inventories had decreased substantially and the speed of response in the

EXHIBIT 4
Homogenized Assembly Sequence, Yanmar (1982)

	One 20-Minute Interval
	Product
	C
	B
	C
	A
	C
	B
	C
	C
	B
	A
	C
	C
	B
	C
	D
	A
Average Days of Inventory	1/2 Day

Source: The Boston Consulting Group: Yanmar Diesel Engines.

shop had increased dramatically. The formerly focused factory is flexible. Which snapshot would the customer prefer?

The operation at Yanmar in 1982 was visibly more flexible but was it economical? What about the diseconomies of flexibility and small batch sizes? The theory of scale economies suggests that, as batch sizes and production runs increase, costs decrease and productivity rises. Furthermore, conventional wisdom indicates that product-line complexity is costly. The experience at Yanmar flies in the face of those dicta. In fact, while the manufacturing transformation itself is impressive, the change in production economies is more remarkable, as shown in Exhibit 5.

The data in Exhibit 5 indicate that, over a six-year period, product-line breadth was quadrupled and *productivity increased*, batch sizes shrunk to the minimum and *costs decreased*! To the

EXHIBIT 5
Change in Production Economics, Yanmar Diesel

	1976	1983
Productivity index	100	191
Cost index	100	44–72
Breakeven as percent of capacity	80	50
Complexity index	100	370

Source: The Boston Consulting Group.

competitor wedded to a focus strategy, these numbers are terrifying.

How are these changes possible? The Yanmar example illustrates that flexible manufacturing offers an escape hatch from the variety trap. A closer look reveals that JIT systems are the key to achieving this manufacturing capability. Just-In-Time and flexible production systems go hand-in-hand. It is useful to step back here and examine the origins of JIT.

THE ORIGINS OF JUST-IN-TIME—A TALE OF TWO COMPANIES

Although JIT is widely considered to be a revolution in manufacturing, a careful examination of its origins leads to the startling conclusion that the system may have been developed as a response to market needs. To be competitive in global markets, firms must possess distinctive attributes: world-class quality, product-line breadth, and rapid response to customer needs. Previous sections demonstrated that conventional manufacturing processes cannot meet these objectives and, at the same time, employ long production runs desired for scale economies. The inflexibility and slow response of large-batch manufacturing created pressure for the development of a new type of system and what emerged was JIT. The old dictum that "Necessity is the mother of invention" aptly applies to JIT.

The JIT production systems evolved over time in response to external pressures. Development of those systems was not guided by a blueprint or master plan. Taiichi Ohno of Toyota, generally recognized as the father of JIT, confirmed this point: "I think that we can only understand how all of these pieces fit together in hindsight. I do not believe that Toyota was guided by the seven principles of JIT in their development."[10]

This writer agrees with Ohno's assertion that the origins of JIT can be understood only through the rearview mirror. While the version of those origins presented here uses many of the sources used by other authors, it also presents a different perspective on JIT. A view from a different angle is necessary to understand JIT in a wide context, a context that fosters transferring the technology to other parts of the value-delivery chain.

The evolution of JIT can be traced through two companies' experiences and their reactions to market pressures: an American company, Ford Motor Company, and a Japanese company, Toyota Motor Company. Although Taiichi Ohno and his colleagues at Toyota are credited with the parentage of JIT, they acknowledge that the seeds of the system came from the United States. Many ideas for modern production systems were germinated by Henry Ford in the 1920s. Ford almost had it, but lacked one essential element: flexibility in manufacturing. Ford's development of the concepts for the modern assembly-line production system laid the groundwork for the JIT system that emerged in the 1970s.

Henry Ford was a visionary and, in many ways, far ahead of his time. He recognized a tremendous market opportunity for low-cost transportation and developed the world's most efficient and timely system for producing cars. The giant, fully integrated River Rouge plant in Detroit was devoted to producing a single product: the Model T. In 1924, boats carrying iron ore docked at the plant and unloaded. Ore was smelted and processed into steel which was forged into engine blocks and springs and axles. These parts were machined, sent to the assembly plant, and assembled into cars. According to published

---

[10]Taiichi Ohno, "The Origin of Toyota Production System and Kanban System," *Proceedings of the International Conference on Productivity and Quality Improvement*, Tokyo, 1982.

reports, this process took about 81 hours—essentially a JIT plant. Thus, 70 years ago, Ford was the world's leading time-based competitor.[11]

This was world-class manufacturing in the 1920s. If Ford Motor Company did it then, why can't manufacturers do it today? Ford's River Rouge plant, however, was a focused, dedicated facility devoted to the production of a single product in a single color. Manufacturing lacked flexibility and therein lay the seeds of its ultimate failure. As consumer tastes changed, so did the demand for cars with improved features, new technology, and more colors. Ford's rigid and inflexible system proved to be incapable of responding rapidly enough to customer demands. It was doomed to failure. Ford missed one crucial piece of the puzzle: flexibility.

Ford had quick response, but not variety. In the late 1970s, Toyota pulled it all together by developing a system with both speed and flexibility for responding to market changes. According to Ohno, Toyota reacted to a crisis.[12] In fact, manufacturing itself reacted to a chain of events that began with a crisis in marketing and trickled down to manufacturing. The crisis grew out of a need for product-line variety and manufacturing's ability to produce it quickly.

To succeed in the economic environment of post-war Japan, Toyota needed diversified, small-quantity production. Although it had the capacity to produce 1,000 vehicles per month, the market demanded that production be spread over a mix of products: four-ton trucks, one-ton trucks, small passenger cars, and so on. American methods for low-cost mass production would not satisfy Toyota's needs. That is, a rigid "focused factory" would not do. A simple transfer of the Model T assembly techniques was not Toyota's answer. To compete in the fiercely competitive Japanese market and, ultimately, in global markets, Toyota needed flexibility. Today, most of the world's automakers wish that Toyota had adopted a focused-production strategy. Al-

---

[11]Henry Ford, *Today and Tomorrow* (Garden City, NY: Doubleday Page & Company, 1926).
[12]Taiichi Ohno, *Proceedings, Productivity and Quality Improvement.*

though the development of a similar production system would have occurred eventually, Detroit may have had a longer reign as king of the automotive hill.

Instead, Toyota eliminated the complexity problem through a direct, frontal attack. The company addressed the cause of the problem—setup, or changeover, time—rather than the symptoms. The solution was remarkably simple: Ohno hypothesized that setup times were wasteful; by eliminating the waste, the cost of variety would be minimized. Setup-time reduction is the key because it represents the most direct attack on waste and the cost of variety. It makes smaller batch sizes economical, thereby solving the complexity problem. Fast changeover, in turn, becomes the catalyst for lead-time reduction which provides quick response to fluctuations in market demand. Reduced setup times alone, however, did not produce the JIT systems that are widespread today. This step is merely the first (albeit the most important) in the full implementation of JIT. Ohno has said that "Ideal Just-In-Time production is only a dream unless setup time is reduced."[13]

Reducing changeover time and converting to smaller batch sizes unleashes a raft of new problems—such as quality control and material flow (addressed later in this chapter)—the solution to which creates higher velocity in the production process. Toyota's achievement of high velocity yielded a production system that is far more responsive to market demands than any prior system. Conversion problems at Toyota were difficult; its system underwent years of development. Followers, including Yanmar Diesel, however, mimicked the conversion process in a fraction of the time. A growing number of U.S. firms, led by Hewlett-Packard, John Deere, and Omark Industries, have successfully converted to JIT.[14]

The steps that Toyota took to develop flexible, JIT manufacturing warrant closer examination. Similar approaches to non-manufacturing operations beyond the factory's walls can have

---

[13]Ibid.

[14]Richard J. Schonberger, *World Class Manufacturing: The Lessons of Simplicity Applied* (New York: The Free Press, 1986).

powerful effects on a company's value-delivery chain. By transferring the technology, it is possible to achieve dramatic reductions in new-product development, engineering, and the paper flow processes that characterize customer service operations. Later chapters of this book examine how to apply JIT principles to other parts of the chain. These chapters explain that the goal is not time reduction in manufacturing alone, but compression of the total time in the chain that delivers the product or service to the customer.

INITIAL STEPS IN JUST-IN-TIME'S DEVELOPMENT

The First Step: Locating Waste in the Operation

In its campaign against complexity, Toyota was guided by a philosophy that waste in any process is both undesirable and unnecessary. Although opinions about JIT differ, virtual unanimity prevails concerning the definition of waste: Anything that does not add value to a product or service constitutes waste. This definition has been chanted in countless seminars, conferences, and publications until all manufacturing managers know it by heart. Waste is why it takes four weeks to process a customer's order that is comprised of only 15 to 20 minutes of value-added time.

The old adage, "Waste not, want not" is apt. To satisfy the customers' wants, value must be added. Customers presumably pay for value. A productive way to identify and eradicate waste, therefore, is to focus on what adds value. Everything else is extraneous.

How is value added to a product or service? How is it measured? This is an elusive concept to capture. Some authorities maintain that only activities that physically change or transform the product add value. This is correct, for the most part, in manufacturing, in which value is added by changing the form of parts or assembling them. However, moving a product can, under certain circumstances, also add value. For example, distributors can add value only by moving a product from the producer closer to the customer. In banking, value is added only when money is moved from one place or account to another. A financial intermediary's basic function is to move money from a supplier to the customer and, sometimes, serve as a broker and locate the supplier. Extraneous time in this process equates to time in which the bank must hold money in inventory as a non-earning asset. Waste in a bank is clearly time. Eliminating that waste benefits the bank in several ways. First, the bank, through faster transactions, can move closer to its customers and serve them better. Second, the removal of waste creates a process that is not only faster, but also more economical. The customer is happier and so are the stockholders.

What emerges is this: Value is added by changing the form of something or by moving it closer to the customer. Activities that do neither are waste, and so is inactivity. In the product development process, new designs that languish on paper or in electronic form inside a computer awaiting further work or review are waste. In order processing, a batch of customer orders awaiting approval or a credit check represents waste. These inert states are merely idle time because nothing is being done to add value.

To add value and eliminate waste within a manufacturing operation, waste first must be located and identified. The sources of waste are found by close observation of the specific process and all its components. Each step of the operation, in sequence, must be identified, charted, and categorized as adding or not adding value. In a typical process, most of the operations, and up to 95 percent of the elapsed time, will prove to be effort that adds no value. For example, inspections add no value; neither does waiting in queues, moving from one work station to another, adjustments, or machine setup times.

Surprisingly, many managers do not have intimate knowledge of their operations. They have not worked at each of the stations, used the tools, or made the adjustments, and thus, they have only a general idea of the process. More than "management by walking around" is required to expose waste in an operation. What is required is close observation, questioning, and digging for facts, the culmination of which is a chart that describes each

EXHIBIT 6
Value-Added Time in Boot Manufacturing

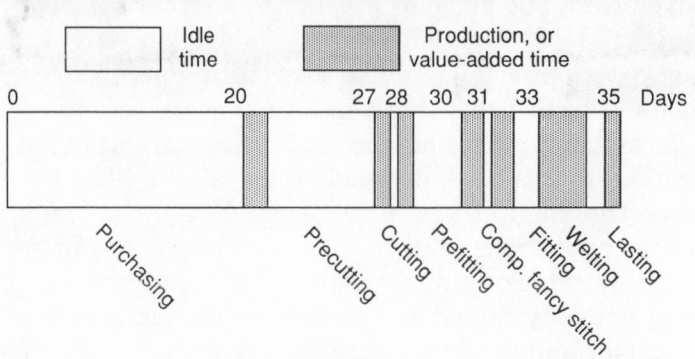

step in the process. The work is hard and tedious. It is not nuclear physics or rocket science, but the results can be explosive.

Management is often shocked by the outcome of value analysis because it brings to view numerous glaring sources of waste. The case of a U.S. manufacturer of western boots is a good illustration. All parties from top management to the shop floor understood the manufacturing process. As diagrammed in Exhibit 6, a simple linear sequence of steps is required: requisitioning of leathers, cutting, prefitting the boots, computerized stitching of uppers, fitting, assembly on the lasts, and welting.

At the time of this analysis by an MBA summer intern, 35 working days or seven full weeks were needed to complete an order for a batch of boots. This time frame was the industry standard. Yet the value analysis revealed that, during most of that 35-day period, no value was added to the boots. The wasted time is illustrated by the white void on the diagram in Exhibit 6. Only the shaded area denotes time when work was performed that added value to the product.

How long would it take to make a pair of boots with no delays and no waste? Experiments indicate that a pair of boots can be made in less than a day. Although some idle time accrues while the boots are being shaped on the last, the boots still could be finished in a single shift. What happens to the boots during the remaining 34 days? They are clogging up the flow by sitting in WIP inventory. This WIP forms a long pipeline through which all the orders must flow. Not only does WIP increase the manufacturing lead time, it also creates a host of other problems. Quality is affected because WIP increases the time between production of a defective part and its detection. Time delays block quality improvement programs by making isolation of the source of defects more difficult. In addition, scheduling becomes unmanageable because WIP increases lead times and lead-time variability, making it impossible to predict when jobs will reach specific work stations.

For the boot manufacturer, the value-analysis jump started a management review that soon excised two weeks from the sequence. Initial changes recovered most of the time from order processing and procurement. Subsequent steps also compressed time in manufacturing. The value analysis indicated the need for modular production lines to partition the flow and increase the fraction of time spent in actual processing operations. Modular production decreased the time that orders sat in queues of WIP inventory.

Changeover times are a major source of waste in conventional batch manufacturing. They add no value to the product. In fact, a setup on a bottleneck machine is a shut-off valve on the entire factory. Effectively, output is zero while the bottleneck equipment undergoes a changeover. The philosophy underlying JIT can be seen quite clearly here: The cost of variety in the factory is the cost of changeovers; eliminate the changeovers and the costs of variety vanish automatically. Eliminate the waste! The concept is deceptively simple.

An operations analysis carried out at Yanmar Diesel in 1976 would have shown that no value was being added while changing over from one model to another. At Yanmar, Toyota, and other firms, setup reduction—in the initial step in the evolution of a JIT process—was the logical consequence of a desire to eradicate waste from the operation.

Step Two: Reducing Setup Time

Setup time reductions are the time-compression process in microcosm. A maximum-speed setup, or changeover, demands a precisely orchestrated sequence of activities, many of which must be carried out simultaneously. Equivalent skills are needed to remove time in any process. Understanding how to reduce setup times, therefore, is the first lesson in the education of a time-based competitor.

A precise, scientific method for reducing setup times exists and is well documented. Although the techniques were sharpened through years of experience at Toyota and other Japanese firms, a trip to Japan to learn how to reduce setup times is unnecessary. The basic how-to guide was written by Shigeo Shingo, who worked with Taiichi Ohno at Toyota. Shingo's seminal work, *A Revolution in Manufacturing: The SMED System*, outlines a procedure that has produced dramatic time reductions in hundreds of cases.[15] SMED is an acronym for Single Minute Exchange of Die. Pursuit of the objective of die changes under one minute led Shingo and his colleagues to discover a procedure that is generally applicable to all setups.

Fortunately, the technology for reducing setups is culture-free and is readily transferred to U.S. manufacturing. Numerous setup-reduction experts have helped American manufacturers refine and apply Shingo's techniques. One of these experts, Ed Hay of Rath & Strong, claims in his book, *The Just-In-Time Breakthrough*, that "The setup time on any piece of equipment can be reduced by 75 percent without major expense."[16] Without extensive automation or elaborate robots, the setup time on any machine can be cut to one fourth its former length.

In his writings, Shingo concentrated on the scientific technique of setup time reduction. Others, such as Hay, embellished the procedure by emphasizing the importance of team-building concepts. Hay stresses the critical role played by the team leader and the importance of team formation. To be successful in this effort, the entire team must be motivated to pursue the single goal of minimizing machine downtime between making the last part on one batch to making the first good part on the next batch.

---

[15]Shigeo Shingo, *A Revolution in Manufacturing: The SMED System* (Cambridge, Mass: Productivity, Inc., 1985).

[16]Edward J. Hay, *The Just-In-Time Breakthrough* (New York: John Wiley & Sons, 1988).

EXHIBIT 7
Setup Time Reduction Activity Sequence

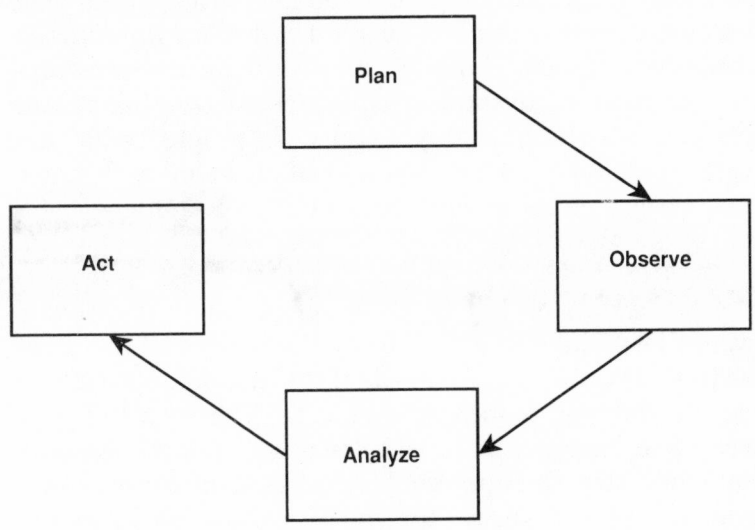

The stages in a setup time-reduction project are similar to those of any scientific investigation: plan, observe, analyze, then act. Those familiar with the "Deming Wheel" will be struck by the similarity between the steps that Deming advocates for quality improvement and Shingo's plan for setup time reduction. In fact, Shingo's steps are workable for time-compressing any activity. These steps will be our guide for the time-compression techniques described in subsequent chapters. Shingo's steps are diagrammed in Exhibit 7.

In the planning stage, the setup team is formed and the goals for the team are specified. The goal must be to obtain the minimum machine downtime. It cannot be to reduce cost or to save labor because these considerations will deflect the team from its mission of time compression. Certainly, management must enthusiastically support the goal of the setup-reduction team; in fact, obtaining top management's support and approval must precede the formation of the team. Essential team members include the employees who perform the setup and a technical adviser, such as a process engineer. The team leader plays a critical role. The leader must be someone who understands the setup time-reduction process and who can teach and motivate the other members of the team.

In a U.S. facility, team building and motivation of the team can be a daunting task. Unlike the rest of the setup reduction process, team-building techniques that worked in Japan cannot be merely transferred to the United States. Motivating and directing a team requires highly developed "people skills" and what works with Japanese employees is likely to fail with American employees because of cultural differences. Organizational theorists agree that, compared with Japan, team building is more difficult in the United States because U.S. workers tend to be more individualistic and independent. In the United States, workers are less likely to enlist under the banner of the team, whereas the subjugation of the needs of the individual to those of the group is endemic to Japanese culture. The issues of team motivation and leadership are addressed in two later chapters. In Chapter 12, Dan Ciampa, the CEO of Rath & Strong and a colleague of Ed Hay, discusses top management's role in dealing with the "people issues" that arise in this process. In Chapter 6, John Bailey of Honeywell explains how that firm's success in new-product development can be traced to team building.

Once the setup team has been formed, the next step is to observe and record the setup. This process is virtually identical to value analysis carried out to identify the causes of waste in any operation. The objective is to construct a precise operations map of each step in the setup process: the exact sequence of steps, the time required, who performs it, what tools are needed, and so forth. Although the operations map of the setup activities can take many forms, it should be in a form suitable to provide answers to several questions: What is being done? How long does it take? Why is it being done? Who is doing it?

Developing answers to these questions involves interviews with members of the setup team and close observations of the setup procedure. The essential study activity, however, is to make a visual record with videotape. Since the objective is a factual representation of the setup as it normally occurs, it is important to minimize the intrusion of the recording on the normal setup activities. To a certain extent, Hawthorne effects, in which knowledge of being part of an experiment can alter the outcome, are unavoidable. Ed Hay's book outlines some schemes used to ensure that the taping process does not bias the setup activity.[17] Properly filmed, video provides an invaluable record of the setup—every activity sequence can be dissected and each task and worker action can be isolated in slow motion and freeze-frame.

Shingo's basic principles for analyzing the setup activities and acting to modify them are deceptively simple. They are:

1. Classify internal vs. external activities.
2. Complete external activities prior to setup.
3. Convert internal activities to external activities.
4. Install parallel, or simultaneous, activities.
5. Smooth and simplify.

The goal of principles one through three is to reduce to an absolute minimum the number of tasks that must be carried out while the machine is down. This begins with a routine, but powerful, classification process. Using the operations map, the setup team partitions all activities into two groups: internal and external. Internal, or in-line, setup activities can be performed only while the machine is shut down; external, or off-line, tasks can be performed while the machine is running. Off-line activities must be completed prior to shutting the machine down. The clock is ticking when the machine is down, and wasting ticks performing external activities then is costly. In-line activities should be the only ones performed during the critical downtime period. For the majority of unanalyzed setups, large chunks of time can be removed simply by managing the removal of all external activities from the downtime period. Much of the in-line time wasted on external activities, such as validating the work order, searching for the necessary tools and dies, and corraling all the raw materials, can be avoided. Shingo reports, from his years of experience with setups, that from 30 to 50 percent of the setup time reduction can be achieved merely by moving all external activities from in-line to off-line.[18]

---

[17]Ibid.
[18]Shingo, *The SMED System.*

Converting internal activities to off-line takes the team beyond the obvious and often represents a quantum leap in difficulty. In many setups the first internal activities are devoted to machine adjustments. Shifting these adjustments off-line requires ingenious solutions. In molding operations, Shingo observed that molds, once inserted, often had to be brought up to the proper temperature before parts could be produced. By preheating the mold off-line, the internal time formerly devoted to this preparation was largely eliminated.

To reduce the in-line time when the machine is shut down, Shingo recommends parallel, or simultaneous, activities. The operations map of the setup is a valuable tool in deciding which tasks can be carried out simultaneously. Employing simultaneous activities usually involves adding manpower to the setup team because a one-man team lacks the arms and legs required to carry out simultaneous tasks. Shingo also cites the synergy effects gained by adding to the setup team: For instance, two people can carry out changeover tasks in less than half the time needed by a single person. When one person carries out a sequence of tasks, a lot of extraneous walking and movement is involved. Properly planned, much of this movement can be eliminated when two workers divide up the tasks. Thus, a sequence of steps that takes 10 minutes for one worker can be finished in 4 minutes by two workers by eliminating 1 minute of wasted movements from each worker's performance.

Many technical improvements suggested by Shingo and others minimize the time required to make adjustments to machines and thereby reduce in-line time for setups. These include the elimination of nuts and bolts, replacing them with clamps, springs, and cams. Other techniques are the use of U-shaped washers rather than rings; centering devices and blocks instead of elaborate machine adjustments. For the most part, these technical solutions apply only to machine setups and are not applicable to time reduction in general. For this reason, these solutions, while significant, lie outside the scope of this book; the interested reader is referred to the books by Shingo, Hay, and others.

Simplifying and smoothing the in-line activities squeezes out the remaining waste in the setup. Teamwork is essential here to perpetuate parallel activities; the transition from one task to another should be seamless. Coordination at this level requires practice, study, and rehearsal akin to directing a top-flight symphony orchestra. Like the orchestra, the changeover takes time to perfect: Some of the setup reduction cases reported by Shingo took more than a year to complete.

One of the unlikeliest, but best, places to see a fast, well-orchestrated changeover team is at an automobile race. Simply observe the pit crew at an Indy race or a "Winston Cup" stock-car race to see all the elements of a time-compressed setup. In a typical pit stop the car races in, takes on four new tires, a full tank of gas, a quick cleansing of the windshield, and, barring a major screw-up, is back to racing, almost always in less time than a 30-second commercial break. The pit team swarms over the car in such frenetic activity that close observation is required to document each step of the operation.

As with a rapid machine changeover, a world-class pit crew reaches top speed after years of study, analysis, and practice. First, the team is assembled and trained under the guidance of a grizzled crew chief. According to Jeff Hammond, pit crew chief for NASCAR driver Darrell Waltrip, "A crew chief has to be like a coach, he must motivate and inspire his people to do their best."[19] Videotape is used to break down every step in the sequence, allowing the team to examine the minutest actions to find simpler, smoother ways to do them. All the external tasks are performed off-line: tires are ready, tools are handy, the fast-flow gas tank is full and awaiting the car. There is no searching for tools or materials when the car comes in to the pit. Every move of the pit crew has been rehearsed until, at race time, the entire sequence is orchestrated flawlessly. As with setups in the factory, speed comes with practice and analysis over a long time. NASCAR data show (see Exhibit 8) that the approximate time for a racing pit stop has been reduced by 90 percent from the 1950s to today: from four minutes in 1950 to about 20 seconds in 1989.[20]

---

[19]Jerry Potter, "Crew's Speed Just as Critical as Car's," *USA Today*, February 14, 1990, p. C3.

[20]Ibid.

EXHIBIT 8

Improvements in Pit Stop Times (Time for a NASCAR Pit to Change Four Tires and Add 22 Gallons of Fuel)

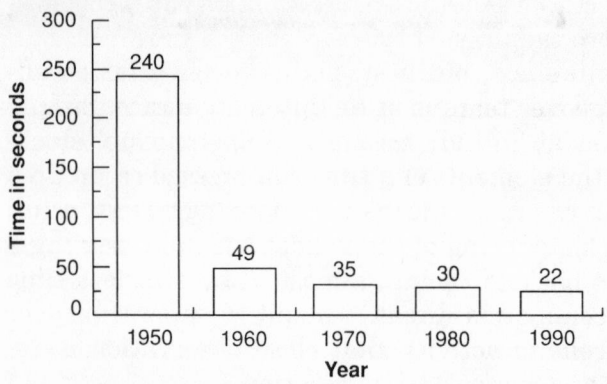

Source: *NASCAR; USA Today.*

Setups on the factory floor should be handled like pit stops. In fact, the machine is in a race to produce parts, and the goal is to minimize the time needed to change over from producing one type of part to another. The changeover team needs a crew chief, and every member of the team must spring into action as soon as the machine has produced the last part of the prior batch. Tools must be ready; all external activities must be completed. Each step in the actual changeover should be coordinated to get the machine off and running again on the next part.

To see how operations analysis and Shingo's principles of setup-time reduction are applied, consider the setup for the leather-cutting step that initiates the manufacture of boots. In cutting, all the leather pieces used to assemble an order for a type of boot in a single size are cut with dies from sheets of tanned hides. The results of observations made prior to the reduction effort are shown in Exhibit 9. Clearly, a substantial fraction of the total setup time is devoted to external activity: obtaining material, checking the order, and selecting the proper dies. Much of the setup analyzed was devoted to external activity because, by tradition, a single operator was responsible for

EXHIBIT 9

Setup Time for Leather-Cutting Operation in Boot Manufacturing
(Before Analysis)

Activity	External (E) or internal (I)	Average activity times (min.)	Elapsed time (min.)
Order tickets verified with leather inventory	E	1	1
Leather requested	E	2.8	3.8
Leather placed on cart	E	3.4	7.2
Cutter retrieves leather	E	6.1	13.3
Cutter retrieves dies	E	2.1	15.4
Die replacement	I	4.1	19.5
Cutter returns old die	E	1.6	21.1
Leather stacked for cutting	I	1.8	22.9
Cutting begins			

all the activities in the cutting process, including the setup. Skilled cutting machine operators, therefore, carried out a number of relatively unskilled tasks relating to material handling and order verification. Approximately 75 percent of the setup time was devoted to external, off-line activities.

EXHIBIT 10
Setup Time for Leather-Cutting Operation in Boot Manufacturing
(After Analysis)

Activity	External (E) or internal (I)	Average activity times (min.)	Elapsed time (min.)
Orders verified off-line before shift begins	E	Off-line	0
Leather delivered to cutter	E	Off-line	0
Dies delivered to cutter	E	Off-line	0
Verification of materials and dies	E	0.5	0.5
Die replacement (helper returns old dies)	I	4.1	4.6
Leather stacked for cutting	I	1.8	6.4
Cutting begins			

Exhibit 10 shows the result of taking off-line most of the external activities entailed in the leather-cutting operation. The changeover time was reduced from 23 minutes to a little over 6 minutes. That is, about 72 percent of the setup time was removed by planning all the activities that must actually precede the cutting operation and by using cheaper, unskilled labor to perform these activities. At this point, nothing has been done to simplify and smooth the actual setup of the cutting operation. Still more could be, and will be, done to trim the time spent on the internal activities that govern this setup process. Major increases in the productivity of cutting machines and machine op-

EXHIBIT 11
Setup Time Reduction Examples

Firm	Operation	Results	
Toyota	Machining fittings	Before	32 min.
		After	7.5 min.
Matsushita	Plastic forming	Before	6 hr.. 40 min.
		After	7.5 min.
Arakawa Autobody	Die change; 500-ton press	Before	27 min.
		After	4.5 min.
Eaton Corp.	Machining center; metal shafts	Before	21 min.
		After	6 min.
Kyoei Kogyo	Die change	Before	4 hr.
		After	10 min.

Source: Interviews; Shigeo Shingo, *A Revolution in Manufacturing: The SMED System*, Productivity, Inc., 1985.

erators, however, have been achieved through application of an operations analysis.

The data shown in Exhibit 11 tend to support Ed Hay's claim that any setup time can be reduced by 75 percent. Moreover, Shingo's techniques work on both sides of the Pacific; some U.S. firms have attacked long setup times and achieved time reductions comparable to those of the leading Japanese firms.

Reducing Setups Creates New Problems

Setup-time reductions are the catalyst for what is essentially a process of time and space compression. Once the economic barriers to small-batch production are eliminated, the manufacturing process is poised for an incredible transformation. It can now be responsive to customer demands; products no longer move ponderously through the shop. Reducing changeover times is just the first step in a journey—a journey of continual improvement (or Kaizen). According to Shingo, "Like priming powder, the effects [of setup-time reduction] touch off other improvement activities throughout the company."[21] Reduced setups

[21]Shingo, *The SMED System*.

produce immediate benefits but, significantly, that improvement uncovers other problems that must be solved. To envision why this occurs, the Japanese invoke the "water and rocks" analogy. This picture has achieved cliché status through its numerous appearances in the JIT literature. Though trite, the cliché is still true: The imagery accurately depicts the event sequence that occurs. In this familiar analogy, the flow of products across the factory floor is analogous to a stream of water. The water level of the stream depicts the WIP inventory level on the floor. In conventional manufacturing, high levels of WIP and large production batches obscure underlying problems in the way that a stream's high water levels hide the rocks on the bottom. The "rocks" on the factory floor include quality, scheduling, machine manning, and layout problems. The goal of JIT is to diminish the "water" (inventory) level in production and increase the rate of flow; setup-time reductions are the first step. As the inventory level is diminished, however, long-hidden problems emerge to block further reductions in time and space.

The "water and rocks" analogy is useful, at this point, as an antidote to managerial panic. Given an initial goal of reducing changeover times, management can, and usually does, succeed. Frustration, however, follows success in the JIT campaign because the resultant problems that surface appear to be caused by JIT. In point of fact, the problems preexisted and became apparent only because of the reduction in inventory achieved through JIT. Now, these problems demand redress. This is not only the time of utmost frustration in JIT implementation, but also the time of greatest opportunity. Setup-time reduction and small batch sizes create an environment for change in which the production process can become a flexible, timely competitive weapon. Exhibit 12 shows the manufacturing evolution that is catalyzed by setup-time reduction. Faster setups make smaller batches possible and reduce the cost of product variety (see, for example, Yanmar's production sequence in Exhibit 4). However, with smaller batches and lower inventory levels, problems surface most notably in two critical areas—quality control and material flow.

EXHIBIT 12
JIT: Evolution of Process

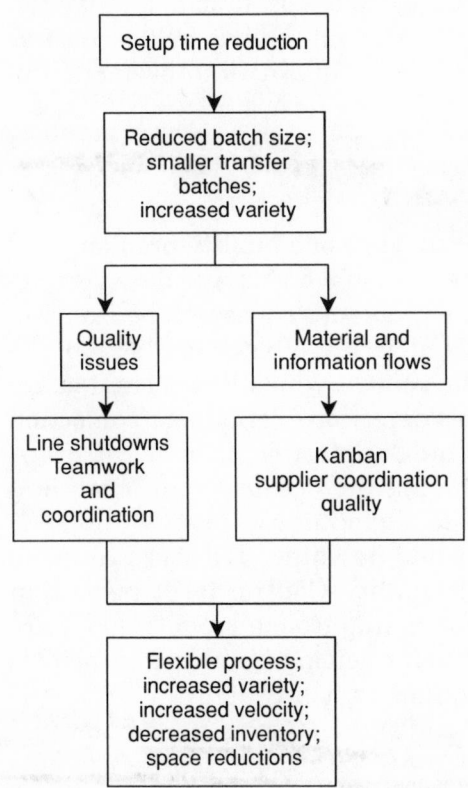

Quality Problems

In conventional batch manufacturing, large batch sizes ameliorate the problems of defective parts; in JIT, small batches expose and amplify quality problems. A few defective parts in a large batch are merely an annoyance ("We can always produce a few more" in the long production run), and the scrap rate, as a percentage, is low. In a small lot, no cushion is built in because producing additional units is wasteful. A scrap percentage can jump into double digits upon production of the first defective part. In a batch size of one, for example, a defective part immediately discovered at the next work station can halt the production line, alerting everyone to the quality problem. Small-lot production, therefore, sparks a cycle of continuous quality improvement because the havoc created on the line by a few defective parts draws management into the quality improvement process.

In JIT, synchronization of adjacent work stations implies a short interval of time between production of a defective part and its discovery at the next station. Prompt discovery of defects intensifies recognition of the occurrence of a quality problem, and reducing the time between creation of a defect and discovery increases the likelihood that the problem's cause will be isolated. Quality begins at the source, and, in large-batch production, the source of a defect and even the time at which it was created are lost. Defect correction requires detection. Permanent correction of quality problems requires quick detection of their causes. A large batch is the haystack hiding the needle; small-batch production reduces the haystack to a handful of straw.

Since quality inspections add no value, JIT seeks to make them irrelevant. Traditional Quality Control (QC) inspection techniques based on random sampling from a production lot are questionable in any manufacturing setting (sampling principles are derived from the assumption that any item in the lot is equally likely to be defective; in manufacturing, the assumption has been invalidated empirically because the first and last units in a lot have the highest probability of being defective due to machine settings and tooling wear, respectively). As batch sizes are reduced in JIT, QC sampling procedures lose any remaining validity. In a short production run, each unit produced must conform to specifications; the operator does not have the luxury of "burning a few parts" at the beginning of the batch to get the machine settings right (this is not to say that the practice should be encouraged in large-batch production). In fact, the goal of setup-reduction activity is to eliminate machine adjustments so that the first part produced is as good as the remainder. If the first part is on spec, then chances are good that the succeeding units are right. As batch sizes are reduced over time, approaching one asymptotically, unit one achieves unique significance.

Therefore, JIT does not just reveal quality problems; it creates an environment that promotes their solution. In the case of the boot company described previously, for instance, managers observed that, as WIP inventories and throughput times were reduced, quality and rework problems that had been obscured became apparent. The quality problems, though always present, had gone unnoticed, and the rework contributed to longer lead times. A companywide focus on quality improvement came as a direct consequence of JIT implementation.

Material Flow Problems

Reduced batch sizes and lowered inventory levels, achieved through setup-time reductions, create problems in the flow of material between work centers. Conventional large-batch manufacturing creates stacks of WIP inventories that isolate and decouple work stations. Work stations are like islands surrounded by oceans of inventory. Machines do not have to be linked and synchronized because large buffers make precise control unnecessary. The distance between work stations tends to be great, and the communication between them tends to be small. Parts are stored and moved in bins, sometimes over large distances; management of material handling often is the responsibility of a separate department.

Small batch sizes and frequent changeovers in JIT alter all this. The oceans of inventory evaporate, creating space and allowing work stations to be moved closer together for both space and time compression. With space compressed, multimanning of machines becomes possible, spawning a need for new layouts. U-shaped layouts have evolved in this setting so that workers can sequence operations and setups on several different machines with minimum travel time. A significant feature of the U-shaped layout is that workers are also moved closer together, often back-to-back, and coordinate their operations as a team rather than as widely separated links in the production chain.

With work centers linked rather than isolated, synchronized, faster material flow becomes possible. Parts are produced and sent to the next station only when needed, not merely to fill up a bin. Parts are "pulled" through the system by downstream demand rather than "pushed" forward in large batches by dictates of a production schedule. To implement the pull system, simple visual cues were introduced to indicate when a machine needs parts and also control the flow of materials through the system. This innovation is known as the Kanban, or card, that was sent upstream to indicate a demand for parts. The precise form of the Kanbans is unimportant—plastic cards, containers, golf balls, tennis balls—any item can be used to trigger production at a work station. The distinction is that, in JIT, the information flows are local; local control of scheduling is possible in a pull system once the work centers have been linked. In conventional batch manufacturing the factors of space, time, and inventory create insurmountable obstacles to local communication between stations, making centralized computer control necessary.

Supplier Relationships

Material flow and quality problems with suppliers should be addressed *after* the JIT campaign is well underway on the factory floor. That is, once the firm has acted to shrink the size of transfer batches between work stations, to compress layouts, and to eliminate inspections between stations, the next logical step is to extend JIT back to suppliers. This means close coordination with suppliers to ship in smaller batches and to eliminate inbound inspection, incorporating suppliers in a chain of synchronous material flow.

Many U.S. firms—mistakenly seeking an external solution to an internal problem—make frequent supplier shipments the first priority in JIT. The managers of these firms are perhaps motivated by a superficial view of JIT as an inventory reduction tool; moving inventory back in the supply chain, however, rarely reduces either time, cost, or inventory. Just-In-Time consultant Ed Hay states that "My feeling is that you put the burden on suppliers last."[22] That is, first put your own house in order with respect to JIT before asking suppliers to change theirs.

Extending JIT into the supply chain demands a fundamen-

---

[22]Craig Waters, "Why Everybody's Talking About Just-In-Time," *INC.*, March 1984.

tal change in the traditional U.S. buyer-supplier relationship. In a comparative study of U.S., Japanese, and Japanese "transplant" supply relationships, Welch, Jueptner, and Blackburn found that, historically, U.S. manufacturers have adversarial, arms-length relationships with suppliers; supply relationships are maintained through contracts rather than partnerships.[23] In addition, U.S. firms tend to be further from their suppliers, in time and distance, than Japanese firms; American manufacturers have more vendors and less single-sourcing. On the other hand, the leading Japanese manufacturers have strong partnerships with suppliers; their suppliers tend to be closer and fewer in number. In Japan, material is transferred from the supplier as seamlessly as between work stations within the factory: frequently, in small batches, and with no intervening inspection.

American managers, impressed by the supply relationships they see on tours to Japan, have rushed to replicate the Japanese JIT supply chains in the United States. These efforts have frequently backfired due to a legacy of mistrust in the supply chain. The trust needed for a working JIT partnership cannot be created instantly by managerial fiat. Japanese supply chains, like the rest of their JIT processes, have been cultivated like Bonsai trees—shaped and nurtured over years. The Welch et al. study also found that Japanese transplants in Tennessee have encountered similar obstacles when trying to replicate, in the United States, the cooperative supply relationships with U.S. firms that they have with suppliers in Japan. The Japanese transplants discovered that, to develop an effective supply chain with U.S. vendors, the first order of business was to develop trust, a tradition of frequent, timely delivery, and quality. Frustrated by these obstacles, many Japanese transplants in the automotive sector have encouraged their traditional Japanese suppliers to follow them to the United States. In Tennessee and Kentucky, for example, Nissan and Toyota have begun to de-

---

[23]James A. Welch, Peter J. Jueptner, and Joseph Blackburn, "Buyer/Supplier Relations: A Contrast between Japan and the United States," Operations Roundtable Research Report, Owen Graduate School of Management, Vanderbilt University, Nashville, TN, January 1990.

velop JIT supply chains with the same firms that serve them in Japan. The challenge to U.S. parts suppliers is not to "shape up or ship out" but "shape up or not ship."

Employee Involvement

The changes unleashed by reducing setups also create a climate of increased employee involvement. This involvement begins with the team assembled to study and minimize the setup time. In conventional batch manufacturing, inventory buffers decouple work stations and isolate workers. As time and space are compressed in JIT, however, shop floor employees must coordinate their efforts to synchronize the flow of the JIT system and become more involved with the process and with each other's responsibilities. Management priorities now shift to teamwork and to group, rather than individual, incentives. The entire focus for process control becomes local rather than centralized.

The JIT revolution has abetted major revisions in our approach to manufacturing management. One of the large "rocks" exposed by the time and space compression of JIT is the ineptitude of autocratic, top-down styles of management to promote team efforts on the shop floor. For solutions, as with process problems, we look to Japan, and the answers are contained in the volumes that have been written in the past decade on Japanese participative management style and the adaptation of these methods to a "more independent, individualistic" American work force. We will not delve deeply into this topic here, but this is not to slight the importance of these "people issues." In fact, the significance of these issues increases as the concept of time compression is propagated throughout the organization. In Chapter 12, Dan Ciampa of Rath & Strong provides valuable insights into how top management can win the commitment of the total organization to time-based goals.

JUST-IN-TIME: A JOURNEY WITHOUT END

Just-In-Time is a journey without a fixed destination. Like the search for excellence, the process of time compression is literally a journey without end. As batch sizes are reduced, new problems emerge, and as these problems are solved, other problems arise. It is like peeling an onion, except that an onion has a finite number of layers. No firm, not even Toyota, can claim that it has solved all the problems in its production process. An omnipresent problem for management, therefore, is motivating employees to support a project that has no firm completion date.

Although the journey is endless, it is not without rewards. As each new problem is solved, system performance improves; over time inventories are reduced, quality rises, space is saved, productivity increases, and throughput time shrinks. Just-In-Time embodies the concept of continuous improvement—the Japanese word for it is Kaizen. Performance measures that record continuous improvement on the JIT journey are necessary to motivate and rally the troops who otherwise might not have the energy to continue an inexorable battle.

The gains from JIT are worth the effort. Hundreds of firms, in Japan, Europe, and the United States, have followed Toyota's model and have embarked on JIT journeys. These firms have found that it takes time to remove time. At Toyota, for example, reducing setup times lowered throughput time in the plant from 15 to 6 days; layout and material flow changes cut the in-plant time in half; and elimination of virtually all the inventory brought the time down to a single day. However, 30 years of effort were required to go from 15 days to 1 day.[24] For other firms, the documented results, as reported by Schonberger, Hall, Stalk, and others, are also impressive. The small sample of case histories summarized in Exhibit 13 demonstrates that performance improvement with JIT is multidimensional: space, inventory, quality, flexibility, and time all show substantial improvement.

With JIT, the pursuit of time compression does not involve making trade-offs. The conventional wisdom is that, in seeking to make process improvements, we are confronted with Hobson's choices of time versus cost, or time versus quality. Fortunately, this is not the case. Focusing on time is a win-win, rather than a zero-sum, game. As time is squeezed out of the process, the other

---

[24]Shigeo Shingo, *Study of Toyota Production System from an Industrial Engineering Viewpoint* (Tokyo: Japan Management Association, 1981).

EXHIBIT 13
Summaries: Just-in-Time Success Stories*

Firm	Product	Results
Eaton Corp.	Transmission shafts	Flow time reduced from 8 days to 28 minutes; inventory reduced 97%
Omark	Twist drills	Flow time reduced from 3 weeks to 2 days; inventory red. by 92%; rework down 20%
Hutchinson Technology	Computer components	Flow time reduced 50–90%; setup times down 75%; WIP down 40–90%
Hewlett-Packard	Printed circuit boards	Flow time reduced from 15 days to 1.5 days; WIP from $670,000 to $20,000
General Electric	Dishwashers	Flow time from 6 days to 18 hours
Motorola	Pagers	Order to finished goods time reduced from 3 weeks to 2 hours

*Sources include interviews:
Craig Waters, "Why Everybody's Talking About Just-In-Time," *INC.*, March, 1984;
Ed Hay, *The Just-In-Time Breakthrough: Implementing the New Manufacturing Basics* (New York: John Wiley & Sons, 1988);
Richard Schonberger, *World-Class Manufacturing: The Lessons of Simplicity Applied* (New York: Free Press, 1986);
Brian Dumaine, "How Managers Can Succeed Through Speed," *Fortune*, February 13, 1989.

important benefits—cost, quality, and flexibility—are achieved simultaneously.

The goal of JIT should be to remove all the rocks and create a flow process. As Richard Schonberger says, "Simplify and goods will flow like water."[25] Above all, the flow is faster, and speed is the attribute we seek in applying JIT principles to the other parts of the value-delivery chain. The key concepts that drive JIT and create speed have wide applicability. Just-In-Time will be our model as we move outside the factory and investigate time compression in the other functions that speed the product to the customer.

---

[25]Richard J. Schonberger, *Japanese Manufacturing Techniques: Nine Hidden Lessons in Simplicity* (New York: The Free Press, 1982).

CHAPTER 3

THE STRATEGIC VALUE OF TIME

George Stalk, Jr.

Editor's Note: George Stalk, Jr., is vice president and director, The Boston Consulting Group, Inc. He is the coauthor (with Thomas M. Hout) of *Competing against Time* (New York: Free Press, 1990). His article in the *Harvard Business Review*, "Time—The Next Source of Competitive Advantage," was one of the first pieces published on the subject of time-based competition. This paper won the 1989 McKinsey Award for the best paper published in the *Harvard Business Review*. George Stalk is also the coauthor (with James C. Abegglen) of *Kaisha, The Japanese Corporation* (New York: Basic Books, 1985).

Time has become an important means for altering competitive strategies. Although managers are conscious of time, they have given surprisingly limited thought to how time affects cultures and businesses in general.

CULTURES AND BUSINESSES HAVE DIFFERENT SENSES OF TIME

Levine and Wolfe, two University of Rochester professors, studied the differences in sensitivity toward time by the people of various cultures. For example, if one compares the consistency of the times registered by the clocks at the major banks in Japan to that of the clocks at the major banks in Jakarta, the differences are quite dramatic. The clocks in Jakarta are off plus or minus three minutes, but in Japan they are off only plus or minus 30 seconds. It takes the Indonesians 27 seconds on average to cover 100 feet, while the Japanese on average need about 21 seconds. Indeed, anyone who has been to Japan knows that at certain times of the day, even if one does not walk, he or she will be pushed along at 21 seconds per 100 feet. Another test was measuring the time required to purchase the equivalent of a first class stamp with the equivalent of a $5 bill in different countries. The researchers found that in Italy the transaction took 47 seconds but in Japan only 25 seconds were required. The various cultures also define "late" differently. Brazilians did not consider themselves late until they were about 34 minutes beyond the time of their appointments, but Californians considered themselves late when they missed their appointments by only 19 minutes.

Whether one looks at the accuracy of clocks or at the speed of a typical pedestrian's trip down a sidewalk, different cultures have very different paces and senses of time.

Businesses Have Different Senses of Time

Not only do the people of different cultures have different senses of time, but business executives also think of time differently. Executives' senses of time can be measured in a variety of ways. One telling method is the amount of time required to develop new products. This varies from hourly and daily for television news programs and newspapers to decades for pharmaceutical companies or defense electronics contractors. The executives of these companies have different senses of time. If a person is 5 or 10 minutes late for a television news show, there is pandemonium; however, if a meeting at a pharmaceutical company is canceled, it may not be rescheduled for six weeks.

Time Is the New Key to Success

Executives of businesses in industries with limited sensitivity to time have a tremendous competitive opportunity, especially if the executives of the other companies in the industry do not recognize the opportunities inherent in time.

The opportunity presented by time is changing the paradigm for achieving corporate success. Historically, corporations have been successful by providing the most value for the least cost. The new paradigm for corporate success is providing the most value for the lowest cost in the least amount of time. Companies that have adopted this new paradigm for success are emerging as a new generation of competitors that manage and compete in very powerful ways. The managements of these companies make time consumption their critical management and strategic parameter. They report time measurements, as well as as cost, revenue, and price. These are time-based, rather than cost-based competitors.

These companies use enhanced responsiveness to stay close to their customers. Response time is the final measure of how close a company is to its customers—how long must a customer wait from the point at which he or she asks for something until the company delivers it. Time-based competitors direct the benefits of their value-delivery systems toward the most attractive customers, the ones that are willing to pay for responsiveness, fast delivery, and choice. The unattractive customers—those who will not pay for responsiveness and instead are willing to wait for the best price—are left for the competition. Time-based competitors grow faster, and with higher profits, than their cost-based competitors. Finally, time-based competitors set the pace of innovation in their industries. In every industry where a time-based competitor has emerged, this company has become the technological leader.

Examples of Time-Based Competitors

Wal-Mart, a discount store, has an 80 percent response advantage over the typical company in the discount industry as a whole. Each stock item in a Wal-Mart store is replenished twice a week. The industry average is once every two weeks. This enables Wal-Mart not only to turn inventories faster but also to change the complexion of the stores to match demand and to offer more product breadth with lower investment in inventory. These advantages are fueling Wal-Mart's growth, enabling the company to grow three times faster than the industry and to be twice as profitable as the industry average.

Atlas Door, an industrial door manufacturer, has a delivery time advantage that is 66 percent faster than its principal competitor—the Overhead Door Corporation. Today, Atlas Door is the leading producer of industrial doors in North America. The company is growing three times faster than the industry and is five times more profitable than the industry average. In contrast to Atlas, the management of the Overhead Door Corporation, who sees only 2 percent return on investment, believes that the industrial door industry is a commodity business in which no company can make money.

Wilson Art is the brand name for a company called Ralph Wilson Plastics. Ralph Wilson Plastics (RWP) makes decorative laminates—commonly known as Formica. Formica created the decorative laminates business, but RWP now dominates it. RWP delivers 75 percent faster than Formica and it can manufacture and deliver an out-of stock sheet of laminate in eight days or less. This compares to an industry average of about 33 days. RWP is growing at three times the industry rate and is four times more profitable than Formica. Formica executives have said that there is no way to make money in the laminate business.

Thomasville Furniture is one of three North American furniture manufacturers that have a quick-ship program. Thomasville guarantees that if a customer comes in to buy a piece of Thomasville furniture and the piece is not available, Thomasville will deliver it in 30 days. The industry standard is three months. Thomasville, like the other quick-ship furniture manufacturers, is growing at four times the industry rate and is twice as profitable as the industry average.

Thomasville's growth and profitability are typical of time-based competitors. When a company can provide the key values its customers want three to four times faster than its competitors, the company is likely to grow three times faster than the industry and be twice as profitable as the average firm in that industry.

CHANGING SOURCES OF COMPETITIVE ADVANTAGE

The emergence of time-based competitors is another step in the evolution of competitive business strategy. With each such step, competitive positions are altered as enlightened companies take advantage of their less insightful competitors. Today, time-based competitors are using fast response time and low cost variety to grow faster than their competitors. Leading Japanese and Western companies are compressing the time required to manufacture their products and to move the products through their distribution systems. More importantly, these companies are significantly compressing the time required to develop and introduce new products. These newly developed capabilities reduce costs and enable these companies to increase the breadth of products offered, covering more segments of their markets, and to rapidly increase the technological sophistication of their products.

The Focused Factory

The relationships between cost, variety, and time become clear from a study of the economics of the focused factory (Exhibit 1). Focusing a factory dramatically affects its performance in terms of variety, productivity, costs, and breakeven. Generally, if a company cuts its product line in half the productivity of its labor increases by about 30 percent, its costs decline by 17 percent, and breakeven is reduced. If the product line is cut in half again, or to 25 percent of its original density, productivity improves dramatically, costs decline substantially, and breakeven falls below 50 percent.

Japanese competitors were very effective at exploiting the economics of focus in their drive to penetrate Western markets in the 1960s and 1970s. These Japanese competitors restricted their product lines to half of their Western competitors' product lines. For example, in the late 1960s the large bearing companies in the West experienced intense competition from Japanese competing with narrow product lines focused on high-volume segments, such as bearings for automobile applications. The Japanese built these bearings in very focused factories,

EXHIBIT 1
The Variety Barrier

Variety Index	Productivity	Unit Cost	Breakeven Percent of Capacity
100	100	100	80
50	131	83	61
25	172	69	46

Source: The Boston Consulting Group.

while their Western competitors produced full product lines in much larger factories. Compared to their Western counterparts, the focused factories of the Japanese had high labor productivity and very low costs. Their lower costs enabled the Japanese to undercut the prices of Western companies.

One Western company, SKF, halted the Japanese advance by combining the economics of focus with their established competitive position. SKF reviewed the role of each of its factories and focused each on those products that it was best suited to manufacture. Other products were either placed in another, more suitable factory or were dropped. Today SKF is widely hailed as a company that beat back the Japanese.

From Focused to Flexible Factories

While the economics of focus gave the Japanese a powerful competitive advantage, they soon extended this advantage into a new area. They altered the economics of focus by developing flexible factories. Much of the battering of the U.S. and European automobile producers in the late 1970s is attributable to the brute productivity and cost advantages of those Japanese producers exploiting their version of the flexible factory, first introduced in the Toyota production system.

The Benefits of Flexible Factories

The cost of variety is less in flexible factories than it is in traditional factories. Flexible factories differ from traditional factories on three principal dimensions—lot size, flow patterns, and scheduling.

Along the dimension of lot size, the traditional approach to factory management in the West has been to maximize production runs. Japanese companies try to shorten their production runs. Indeed, for many Japanese companies the goal is to achieve run lengths of one unit. Reduced run lengths mean more frequent production of the complete mix of products and faster response to customers.

The layouts of traditional factories differ from those of flexible factories. Traditional factories are often organized by process technology centers. Examples include the shearing, punching, and braking departments for metal goods and the stuffing, wave soldering, testing, assembly, and packing departments for electronic assemblies. Production parts are moved from one process technology center to another. Time is consumed as parts wait to be moved, are moved, and wait to be used in the next step.

Flexible factories are organized by product. As many functions as needed to manufacture a component or product are brought as close together as possible to minimize the handling and moving of parts. A part moves from one activity to the next with no or very short delays. Parts are not piled and repiled. In product-oriented layouts, parts flow quickly through the production process.

The scheduling of traditional factories is complicated by the process center organization. Traditional factories are often centrally scheduled, usually requiring sophisticated MRP and shop floor control systems. These systems direct much of the activity on the floor and feed back to management the results of their decisions. As sophisticated as these systems can be, they still consume time. The floor direction modules are often exercised monthly or weekly. Between exercises, parts wait.

Flexible factories use more local scheduling. More production control decisions are made on the floor without a loop back to management for approval. Local scheduling does not require more capable employees. Quite the opposite is true. The product-oriented layout of the flexible factory means that when a part is started many of the step-to-step movements are automatic and do not require intermediate scheduling. As a result of the changes, flexible factories can lower the costs of variety.

In a factory, costs fall into two categories (Exhibit 2). Some

EXHIBIT 2
Breaking the Variety Barrier

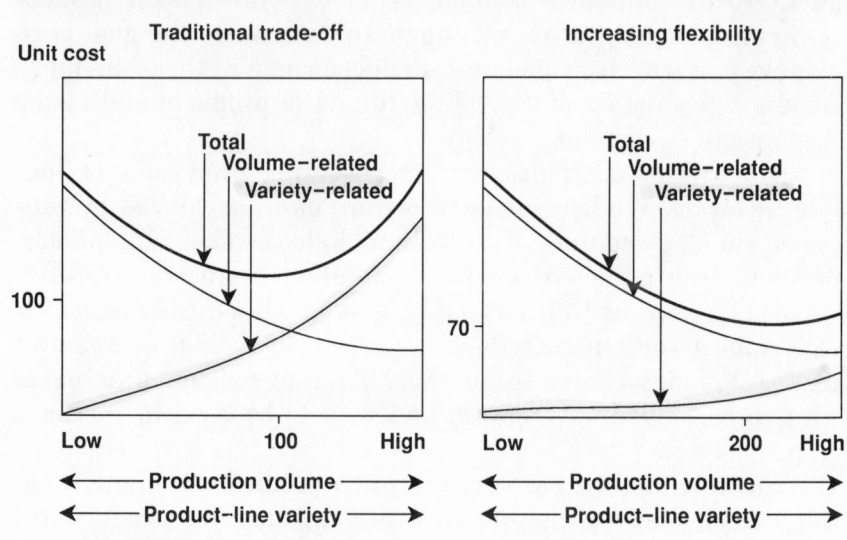

Source: The Boston Consulting Group.

costs are very sensitive to increasing volume or scale. These costs are generally equipment or process related but can include other types of costs. Scale-driven costs decline as volume increases. The typical per unit rate of decline is 15 to 25 percent for every doubling of volume.

The second category of costs is variety-driven costs. These are the costs of complexity. These costs include setup, material handling, inventory, and many of the overhead costs of a factory. Complexity-driven costs increase as variety increases. The typical per unit rate of increase is 20 to 35 percent for every doubling of complexity.

The total cost of manufacturing is the sum of the scale- and complexity-driven costs. Factories have a cost optimum point where the combination of the volume achieved for the variety offered results in the lowest manufacturing costs for that facility. When markets are good, companies tend to creep to increased levels of variety in search of higher volumes even though costs are increasing. When times are tough, companies pare their product lines to lower costs.

When a company makes its factories flexible, the costs of variety are reduced both absolutely and on the margin. This effect is shown in the figure on the right-hand side of Exhibit 2. Scale costs are seldom affected in the medium term by making a factory flexible and, thus, in this schematic are unchanged between the traditional and flexible factories. The cost trace for the variety-driven costs for the flexible factory is lower and increases less with greater variety than the trace for the traditional factory.

When the variety- and scale-driven costs are summed for the flexible factory, the resulting total cost line has a point of minimum total cost that occurs at higher levels of variety and volume than does the minimum point for the traditional factory. A gap emerges between the costs of the traditional factory and those of the flexible factory. This is the cost/variety gap. A company whose factories are flexible can operate with greater variety but with lower costs than competitors whose factories are still operated traditionally.

The cost/variety gap is observable in an increasing number of competitive situations where the Japanese company has flexible factories and its Western counterpart does not. Some data from a recent situation are shown in Exhibit 3. In this situation the U.S. company is the largest manufacturer in the world of this suspension component, with an annual production rate of about 10 million units per year. The U.S. manufacturer is also a very focused producer with only 11 finished part numbers. The principal Japanese competitor is much smaller, with an annual production volume of only 3.5 million units. The Japanese company, with 38 finished part numbers, is also much less focused than the U.S. company.

Despite the fact that compared to the U.S. manufacturer the Japanese company is about one-third the scale and has more than three times the product variety, the total labor productivity of the Japanese company is more than 50 percent higher. Interestingly, the productivity of the Japanese company's direct laborers is not as high as that of the U.S. company. This is to be expected since the Japanese company is at a substantial scale disadvantage. The source of the productivity advantage of the

EXHIBIT 3
Breaking the Variety Barrier: Small Scale, High Variety, and Low Cost
(Automobile Suspension Component)

	U.S. Competitor	Japanese Competitor
Annual unit volume	10	3.5M
Finished part numbers	11	38
Units per employee	40,800	62,600
Employees		
Direct	107	50
Indirect	135	8
Total	242	58
Unit cost for comparable part	$6.73	$3.27

Source: The Boston Consulting Group.

Japanese company is the higher productivities of its overhead employees. At one-third the volume and three times the complexity, the Japanese company has one-twentieth the overhead employees. Flexible factories are easier, not more difficult, to manage than traditional factories.

The high productivity and the willingness of the Japanese company to increase variety has helped it deal with the increasing value of the yen. At 180¥ to the dollar, the Japanese company held a 35 percent cost advantage. At 130¥ to the dollar, the cost advantage of the Japanese company has declined to 15 percent, but sales volume has increased with the addition of new products.

The pattern in Japan of companies becoming more flexible and then experiencing a dramatic increase in labor productivity *while* increasing variety is fairly widespread among the larger companies in Japan. Exhibit 4 shows the improvements in performance of only a few prominent Japanese companies in a five-year period in the late 1970s. These companies improved the productivity of their labor forces by more than 100 percent. In addition, the productivity of the net assets more than doubled. These rates of improvement are far greater than can be explained by their growth in this period. The significance of their

EXHIBIT 4
Typical Performance Improvements (circa 1976–1982)

Company	Product	Factory Labor Productivity	Net Asset Productivity	Product Line Variety
Yanmar	Diesel engines	1.9x	2.0x	3.7x
Hitachi	Refrigeration equipment	1.8x	1.7x	1.3x
Komatsu	Construction equipment	1.8x	1.7x	1.8x
Toyo Kogyo	Cars, trucks	2.4x	1.9x	1.6x
Isuzu	Cars, trucks	2.5x	1.5x	NA
Jidosha Kiki	Brakes	1.9x	n.a.	n.a.
Average		2.0x	1.8x	2.1x

Source: The Boston Consulting Group.

improved performance is even more striking because, as can be seen in the third column, the companies increased the variety offered by 20 to 320 percent and on average 100 percent.

The improvement in the productivities of labor and net assets are of substantial strategic significance alone. Such improvements usually mean a 20 percent reduction in overall costs and growth for half the usual levels of investment. But these Japanese companies are simultaneously reducing costs, expanding variety, and growing.

The fast rates of change are the result of structural changes. These changes enable the organization to execute its basic processes much faster. The thrust of these changes is to *reduce time as a bottleneck.*

Time is a fundamental business performance variable but seldom is its consumption monitored explicitly by management. Almost never is time measured with the precision of the measurements used for sales and costs. Yet, evidence is accumulating that suggests that time should be the parameter of the first order followed by financial parameters.

Why Time Is Important

The pervasive effects of time delays on business can be seen by considering the "planning loop" that is experienced in manufacturing planning. Conventional manufacturing requires long lead times to resolve conflicts between various jobs or activities requiring the same resources. Long lead times require sales forecasts to guide planning. Sales forecasts are always wrong—forecasts by definition are guesses, however informed. The longer the lead times, the lower will be the accuracy of the sales forecast; the greater the forecasting errors, the greater will be the need for safety stocks at all levels. Furthermore, longer lead times increase the chances that unscheduled jobs will crowd out scheduled jobs, resulting in a perceived need for even longer lead times. This is the planning loop.

Often the response of management caught in the planning loop is to ask for better forecasts and longer lead times. However, this is treating the symptom rather than the problem. The way to break the loop is to reduce significantly the consumption of time that creates the need for lead times. If lead times could be reduced to zero, the only sales forecast needed would be the estimate of tomorrow's sales. The chances of accurately estimating the sales for tomorrow, next week, and next month are far greater than those for estimating the sales for the next quarter, six months, or for even 12 months.

Lead times of zero are unrealistic to hope and plan for. However, anything that can be done to keep lead times from increasing or, even better, to substantially reduce them, will reduce the impact of the planning loop on an organization.

A more sophisticated perspective on how time affects an organization's performance is possible using an approach developed by Jay Forrester of M.I.T. At the end of World War II, Professor Forrester developed a technique originally labeled "industrial dynamics" and later relabeled "systems dynamics." Industrial dynamics applies to business problems the control theory that was used to develop shipboard fire-control systems. In an example application of industrial dynamics, drawn from his seminal paper published in the *Harvard Business Review* in 1953, Professor Forrester shows how industrial dynamics can be applied to the dynamic behavior of a fairly simple business system. The article demonstrates how long response times can amplify the effect of change and uncertainty on a business system. The system in Exhibit 5 consists of a factory, a factory warehouse, a distributor's warehouse, and retailers' inventories.

EXHIBIT 5
Time and Business

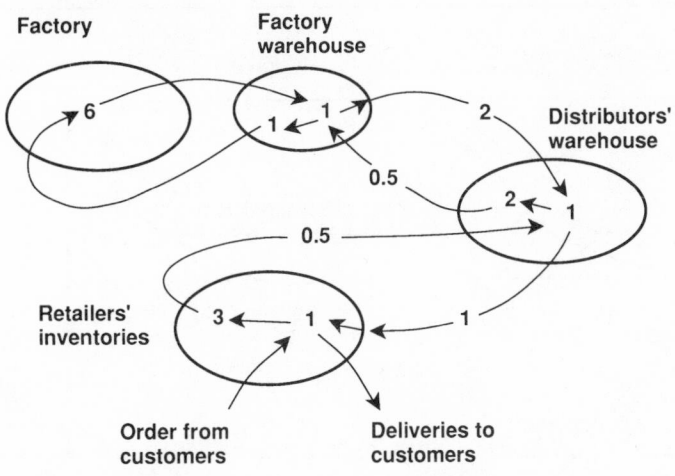

Source: Jay Forrester.

The numbers within the circles are the delays, measured in days, in the flow of information or product. Thus, if there is stock-out at retail, the orders accumulate at the retailer for three days, are in the mail for half a day, are delayed at the distributor for two days, go back into the mail for a half day, need eight days to be processed by the factory and its warehouse, and then the finished product begins its journey back to the retailer. The complete cycle requires 18 days.

This system is very stable dynamically as long as retail demand is unchanging or if demand is known for certain at least 18 days into the future. For other conditions the system will have to respond, and the response will vary with the rate of change and with elapsed time.

Demand does change and is seldom known with certainty. The response of this system to a simple change in demand is shown in Exhibit 6. Demand increases by 10 percent. The factory responds by ramping up production by 40 percent, then cutting production by 30 percent, followed by another production increase. The oscillations occur because the factory receives in-

EXHIBIT 6
Time and Business (Response of Production-Distribution System to a Sudden Increase in Sales)

Order/Production
rate (units/week)

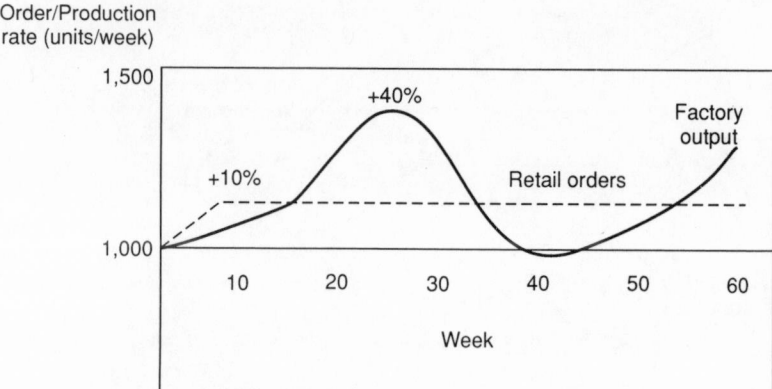

formation many days after the events that generated the information. At each step information is bundled and then passed on. By the time it reaches the factory, it presents a distorted view because reality has changed. It takes almost five years for production to stabilize at a 10 percent higher level.

A more realistic scenario is no growth in yearly retail demand, but random week-to-week demand. The simulated demand and the factory's response are shown in Exhibit 7. The jagged line is the random demand and the smoother line denotes the factory response over a four-year period. The factory settles into a periodic response. The period of the response is approximately 56 weeks or about one year. An executive in this business might believe his business is seasonal based on the demands on his factory. If he attempted to smooth the demand on his factory by advertising and promoting in the troughs and raising prices in the peaks, the oscillations would actually increase.

Many companies have seasonal characteristics and must face seasonal oscillations. These companies have accommodated the demands of the oscillations but with an increased cost of doing business. A business prepared to respond to oscillations has higher costs than one prepared to respond to steady demands.

EXHIBIT 7
Time and Business (Effects of Random Deviations in Retail Sales on Factory Production)

Retail/Production
rate (units/week)

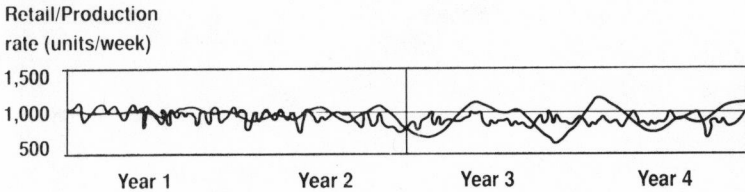

| 1,500 |
| 1,000 |
| 500 |

Year 1   Year 2   Year 3   Year 4

Sometimes, a close examination shows that seasonal businesses are causing a significant portion of the seasonal swings. Examples include luggage, cameras, and automobiles.

Companies and their factories can escape the oscillations only by producing to forecasts or by significantly reducing the delays to the flow of information and product in the system. Producing to forecast is the traditional solution. Significantly reducing delays is the new solution.

If delays to the flow of information and product could be reduced to zero, the factory would be able to respond instantly to changes in demand. No forecasts would be needed. Perhaps zero time delays are too much to strive for. However, if the inherent cycle time in a business is 18 days, it certainly should not be allowed to increase to 24 days because the oscillations only become worse. Any reduction in delay reduces the tendency of the system to oscillate.

In Exhibit 5 the factory and its warehouse have the longest time delays in the entire system. The factory and its warehouse consume 8 of the 18 days of delay in the system. A rule of thumb is that value is added to a product for only one-twentieth of a percent to two and one-half percent of the time it is in the traditional factory. During the rest of the time the product is waiting to receive value. This dead time is the opportunity on which flexible factories are built. Flexible factories consume less time than traditionally managed factories.

Becoming flexible significantly improves the productivity of labor and of assets. Much more impressive, though, are the improvements in response time that result from becoming flexible.

EXHIBIT 8
Typical Improvements in Production Flow Times

| | In Days | | Percent |
	Before	After	Reduction
United States			
Motorcycles	360	<3	99%
Motor controllers	56	7	88
Electric component	24	1	96
Europe			
Twist drills	21	3	86
Gun reloaders	42	3	93
Japan			
Washing machines	15	.1	99

Source: The Boston Consulting Group client assignments and Interviews.

For example, Toyota forced one of its suppliers to become more responsive. With a traditional manufacturing system, this supplier needed 15 days from the time that raw materials arrived at the factory to ship the finished product. When lot sizes were reduced, the time required declined to six days. The required time fell to three days after the factory layout was streamlined to reduce inventory holding points. The elimination of all work in process inventories resulted in response times of one day. This Japanese supplier improved its response time by an order of magnitude.

Many factories are dramatically improving their response time by being more flexible. Examples of improved responsiveness are shown in Exhibit 8. Matsushita reduced the clock time required to manufacture washing machines from 360 hours to just 2 hours. The remaining companies in the exhibit are North American and European, and they have been able to improve the response times of their factories by approximately 90 percent.

Distribution and Sales
In the Forrester example, note that the factory and its warehouse account for about half the time consumed in that system. This is typical of most businesses. The factory generally accounts only for one-third to one-half of the total time consumed.

As in the Forrester example, the sales and distribution systems of most businesses consume an almost equal amount of time and sometimes more.

In Japan, many leading companies had made their factories flexible by the late 1970s but found that their sales and distribution systems were limiting their effectiveness. Reducing the time consumed in the sales and distribution system became the new focus of organizations.

For example, the best factories of the Toyota Motor Manufacturing Company could manufacture a car in less than two days. The sales and distribution system of the Toyota Motor Sales Company required from 15 to 26 days to perform its function of closing the sale, transmitting the order to the factory, and delivering the car to the consumer. By the late 1970s, the engineers at Toyota Motor Manufacturing Company were frustrated that the achievements in their factories were being frittered away in the sales and distribution system of the Toyota Motor Sales Company. Twenty to thirty percent of the cost of a car to a consumer was generated by the distribution and sales function. This was more than it cost Toyota to manufacture the car! If ever there was a company that disliked paying to move product around it was Toyota.

This frustration led to the merger of Toyota Motor Manufacturing and Toyota Motor Sales in the late 1970s. Eighteen months after the merger all the directors of the sales company had been retired and their jobs were either not filled or were filled by executives from the manufacturing company.

The new Toyota developed and implemented a plan to reduce delays and costs in its sales and distribution system. Toyota found that information was being handled in batches, accumulating at one step of the sales and distribution process before being sent to another level. This accumulation consumed time and generated costs.

To speed the flow of information, the accumulation batch size had to be reduced. Toyota developed a computer-based system to tie the salespeople directly to the factory scheduling function, thus bypassing several levels of the sales and distribution system and enabling the modified system to operate with very small information batch sizes.

This new approach to the handling of information in

Toyota's sales and distribution system is expected to reduce inventories and costs. The response time of the system is to be reduced from four to six weeks to two to three weeks across Japan. For the Tokyo and Osaka regions, which account for about two-thirds of the population of Japan, the goal was to improve the responsiveness of the sales and distribution system to two days. By the spring of 1987 the responsiveness of the sales and distribution system had improved to eight days, including the time required to manufacture the vehicle. This is a more than 50 percent reduction in the time consumed in the sales and distribution network.

Toyota's achievements are equivalent to reducing the 18 days in the Forrester example to about 6 or 7 days. Toyota is less affected by random demand. The company can use shorter sales forecasts. Their costs can be less and their customers can be happier.

Expanding Variety

In the early 1980s, the Japanese again moved the leading edge of competition in manufactured goods. The leading Japanese companies began to shift their strategies from exploiting focused product lines made in flexible factories to strategies exploiting broadened product lines. Increased variety became the next leading edge of competitive advantage.

Most Western companies have not yet experienced the effects of a Japanese strategy of increasing variety. This is understandable for two reasons. First, no matter what the culture, competitive innovations are first observed in the home markets. Thus, since few Western companies have an offensive posture in Japan even fewer have directly experienced the exercising of the recently developed Japanese capability to manage increasing variety. Second, most Japanese companies face their Western competitors in the export markets that are usually at the end of very long distribution pipelines. The performance of distribution pipelines deteriorates directly with increasing distances and variety. Many Japanese companies have increased the variety they offer in their home market but have had to restrain the variety sold in the West because of the distribution-related penalties.

The ability to expand variety rapidly can be very powerful. This power can be seen in the variety wars that began to appear in Japan in the late 1970s. In a variety war the aggressive competitor rapidly increases the variety it offers to attract customers away from competitors, to make the product offerings of competitors obsolete, and to force competitors to invest in new products at levels that could be more than they can afford.

A good example of a variety war was observed in the competitive battle that erupted in 1981 between the Honda Motor Company and the Yamaha Motor Company. The battle was popularly known in the Japanese press as the "Honda-Yamaha War" or the "H-Y war."

The H-Y war was triggered by Yamaha. Yamaha had slowly but consistently gained market share over about a 15-year period. Every share point gained by Yamaha was a share point lost by Honda. During this period Honda had diverted its resources from the motorcycle business to its growing automobile business and publicly acknowledged that share loss in the motorcycle business was an unfortunate consequence.

Honda chose to counterattack in 1981 when Yamaha announced that, with the opening of a new factory, Yamaha would be able to fulfill its dream of becoming the world's largest motorcycle manufacturer. This was a position of prestige that was then held by Honda. The attack was launched with the cry, "Yamaha wo tsubusu!" Roughly translated this means, "We will crush, squash, slaughter Yamaha!"

Honda did all the classic things that one would expect in a "no-holds barred" competitive battle. Honda cut prices, it flooded the distribution channels with inventory, and it increased its advertising expenditures. In addition, Honda rapidly increased the rate of change in its product line. In the space of 18 months Honda introduced and retired 113 models. The company had 60 models at the start of the war. Thus, in 18 months Honda essentially replaced its product line twice. Yamaha, which also had about 60 models, was able to make only 37 changes to its product line.

Honda's new product introductions had a devastating effect on Yamaha. Honda managed to increase the fashion awareness for motorcycles in Japan. In addition, Honda raised the technological sophistication of its products by introducing four-valve engines, composites, direct drive, and more. Next to a Honda, a Yamaha motorcycle was perceived to be old and out of date.

People stopped buying Yamahas. To move Yamaha motorcycles, dealers were forced to price them below their costs— something they could not afford to do. At a stroke, the entire field inventory of Yamaha was made obsolete by Honda. At the depth of the war, Yamaha had more than a year and a half's worth of inventory in the field.

After 18 months Yamaha surrendered. In a public announcement the chairman of Yamaha said, "We want to end the H-Y war. It's our fault. Of course there will be competition in the future, but it will be based on a mutual recognition of our respective positions."

Honda remains the world's largest motorcycle producer. Its sales and service network has been severely disrupted by the variety war and requires investment to be stabilized. Strategically, the disruption is not a serious concern. Honda has as much time as it needs to overcome the effects of the war. Yamaha is beaten. Suzuki and Kawasaki know better than to press Honda after seeing what happened to Yamaha. None of the world's remaining motorcycle manufacturers are strong enough to pose a realistic threat to Honda.

The Japanese are just beginning to use variety as a competitive weapon against Western competitors. An example is competition in the diesel engine business. In 1985, a Western company had a diesel engine business that had historically been successful. However, its customers were receiving very low-priced bids from Japanese manufacturers of diesel engines.

The review of the competitive strengths of the Japanese showed that prices could be justified based on estimates of the manufacturing costs of the Japanese. More disturbing, though, was the very different approach to variety of the Japanese. The Western company had been reducing costs by focusing its factories. Its product line had been reduced from five to three basic engines which could be configured to span a very wide range of applications. Their plan was to eliminate another engine by the end of the decade so that the two remaining engines could each be made in its own factory.

The principal Japanese competitor is Hino Motors. Hino is the largest manufacturer of medium- and heavy-duty trucks in Japan. The company is affiliated with Toyota and has adopted the Toyota production system.

Hino offered nine basic engines in the application range for which the Western company offered three engines. In the period that the Western company had reduced its product line from five to three engines, Hino had increased its line from five to nine. In the next five years Hino planned to increase the breadth of its line to 12 engines.

Hino produces its engines in two factories. The two largest displacement engines are made in one factory and the rest are made in Hino's original engine works. The original engine works has an annual production volume equal to that of the largest factory of the Western company. But unlike the factory of the Western company, the Hino factory manufactures seven basic engines instead of two. Further, despite its much greater variety, the Hino factory produces engines at the same level of productivity as the Western factory.

The factories of other Japanese diesel engine manufacturers were also very productive. Komatsu's main engine factory, with an annual production volume equal to that of the largest factory of the Western company, had six times the variety and equivalent labor productivity. At the other extreme was Nissan Diesel. Nissan Diesel is a weak competitor in the trucking industry in Japan. In a new factory built to manufacture medium-duty trucks and the engines they require, the productivity of the engine works was 40 percent higher, even though the overall volume of production was half that of the Western company.

If, compared to the Western company, Hino had had equal variety *and* equal volume, it would most likely have had much higher, rather than equal, productivity compared to that of the factory of the Western company. The fact that Hino does not have a productivity advantage is the result of the company's choice to spend some of that potential advantage on the expansion of variety—an expansion needed to continue to attract customers and grow. Hino was able to more closely tailor its product offerings to the needs of its new customers.

When confronted with the many cases of rapid expansion of

variety by the Japanese, one wonders just how their organizations are structured to accommodate high rates of change. The rate of change exhibited by Honda's expansion of its product line suggests either:

- The development of new products beginning 10 to 15 years before the attack.
- A massive increase in the new-product development and manufacturing resources.
- Structurally different methods for developing, manufacturing, and introducing new products.

The new-product development and introduction cycles of most businesses are also an area where much time can be consumed. In Japan, many leading companies are also pushing the advantages of flexibility into their new-product development and introduction cycles. They are working intensively to reduce the time required in this area. The examples of Japanese companies that are designing and introducing new products very quickly by Western standards are numerous, and include:

- Projection television, where the management of the U.S. company found that the Japanese supplier could develop a new television in one-third the time required by the U.S. organization.
- Custom plastic injection molds, where the Japanese competitors are able to develop the molds in one-third the time required by U.S. competitors at a 30 percent lower cost.
- Automobiles, where the Japanese develop new cars in half the time and with half as many people as required by their American and German competitors.

A specific case is seen in Exhibit 9. The Western company and the Japanese company had been affiliated since the end of the war. The Western company had played a key role in putting the Japanese company into the business of manufacturing marine gears. By the 1980s the Japanese were producing equivalent products at 30 percent lower costs than the Western company was able to produce. More striking though was the comparative speed of the Japanese company's new-product development

EXHIBIT 9
Improving Response Time in New Product Development (Marine Gears)

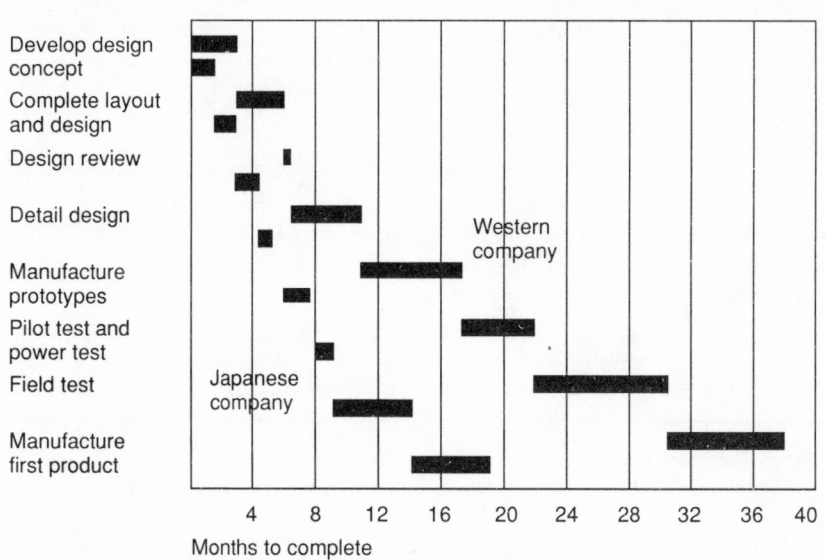

and introduction cycle. For equivalent products the Western company required 34 to 48 months to complete the development and introduction cycle, while the Japanese affiliate could complete the cycle in 12 to 16 months. The Japanese were introducing new products in one-third to one-fourth the time required by its Western affiliate.

Careful inspection of Exhibit 9 yields little insight into how the Japanese company achieved its response advantage. In no one step of the development and introduction cycle—design concept, design engineering, design review, detail design, field test, prototype manufacture, and first production—do the Japanese score a significant advantage. With the exception of the design review step, the Japanese company outperforms its Western affiliate a half step every step of the way.

Structuring New-Product Development

Companies that can introduce new products rapidly are structured differently than their more slowly responding competitors. These companies have applied the same techniques that have

EXHIBIT 10

Contrast in Manufacturing Management (An Analogy)

Factory dimension	Traditional	Flexible
Lot size	Significant improvements planned but fewer projects	Plan for smaller improvements but more of them
Flow pattern	Move through functional centers	Put relevant development resources together
Scheduling	Centrally scheduled	Scheduled within the group
Lead times	100	50
Productivity	100	200
Other	Extensive market research, testing, and deliberation	Market experimentation that, if successful, results in a full roll-out

Source: The Boston Consulting Group.

made their factories so responsive to speed the process of designing and introducing new products.

The structure of a rapid new product design and introduction organization is analogous to the structure of a fast response factory, in terms of lot size, departmental organization, and scheduling (Exhibit 10). Along the dimension of lot size, Western companies plan significant product improvements for each introduction cycle but plan less frequent introductions than do the Japanese. The Japanese plan for comparatively less improvement with each new product introduction but much more frequent introductions of new products. The Western approach is analogous to long production runs and the Japanese approach compares to their desire to have very short run lengths—in their factories and in their product development organizations.

Western companies often organize their new-product development and introduction programs like traditional factories are organized—by functional centers. There are groups for marketing, new-product design, product engineering, and manufacturing. Projects can progress through these functional centers very slowly, just as slowly as products move through process centers on the floor of the traditional factory. The organization of a company that develops and introduces new products rapidly is analogous to the organization of a fast-responding factory. Japanese companies that are organized to design and introduce new products quickly put all relevant development resources together. The group will include marketing, design, manufacturing, and in some cases finance and sales. The participants from these functions are brought together, not on a part-time basis, but generally on a full-time basis. Often, they are physically located at the site where the new product is to be manufactured. Development programs move rapidly through each of the functional activities.

The new-product development and introduction programs of the West must be scheduled for progress to be made. Product managers or program managers are used to plot and track the progress of the program through the functional centers. This is analogous to the use of MRP systems in the traditional factories. In the system structured to introduce new products rapidly, scheduling is done within the group. Start dates for the project are set with milestone dates. Time in between is scheduled by the members of the group. The programs are locally scheduled.

The performance differences similar to those observed for traditional and flexible factories are being seen for the traditional and rapid new-product development and introduction programs. Japanese companies structured to rapidly introduce new products are able to introduce them in about half the time with half the number of people. Integrating these advantages, one observes that, with the same amount of time and the same number of people, the Japanese can introduce four times as many products.

The Leading Edge of Advantage Is Now Shifting to Responsiveness and Innovation

The ability to introduce new products four times faster than Western competitors is a pretty astounding advantage. It is shifting the leading edge of competitive advantage. Starting on the factory floor, leading companies, mostly Japanese, have reduced costs and increased flexibility and variety. These companies then attack the problems of responsiveness and resistance to changes in variety that exist in their sales and distribution systems and in their new-product development and introduction cycles.

As a result of their new abilities, these companies are able to significantly increase their rate of innovation. They are continually increasing the technological sophistication of their products in small increments. After a few years, these incremental innovations create large gaps when compared with the products of traditional companies.

Furthermore, these innovative competitors do not need to decide which old products to eliminate. Their flexible factories can absorb the new products as an expansion of variety. Older products are kept for as long as the market demands them. Of course the new products cannibalize the sales of the old but the time-consuming effort of deciding if the old product should be replaced by the new and then committing to do so is reduced. This is possible because of the flexible factories.

The responsiveness-innovation-variety loop makes possible a new and effective approach for developing markets for new products. In the West, where new-product development and introduction cycles tend to be long, extensive market research, as well as much testing and deliberation, are conducted before a new product is placed in the market. In Japan, where development and introduction cycles are fast, the pattern is one of market experimentation that, if successful, results in full introduction. The contrast is understandable. If development and introduction lead times are very long, a company must try very hard to be sure the introduction will be a success because the long lead times mean that there will not be much of an opportunity to modify the product. When development and introduction lead times are fairly short, a new product can be introduced and then modified quickly as the realities of the market become clearer.

In contrast, the traditional company is under great pressure to make each introduction a success. This company is necessarily risk-averse and slow to innovate. The faster company can risk a "near miss" because it can respond to new developments in the market. This company is less risk-averse and sets the pace of innovation in its industry. Very often, today, the pace is being set by a Japanese company.

One example is the Japanese and American competition in residential air conditioners. The Japanese companies' rate of new-product introduction is four times faster than that of the United States. The technological sophistication of the Japanese products is 7 to 10 years ahead of the U.S. products.

This contrast can be seen in the introduction and innovation history of Mitsubishi Electric, a mid-sized competitor in the Japanese market, but a small company compared to the large U.S. manufacturers. The new-product introductions of Mitsubishi's Japanese competitors are similar and for the most part differ only by the year a change was introduced. The 10-year development history shown in Exhibit 11 is for Mitsubishi's three-horsepower heat pump. A three-horsepower heat pump is the mainstream product in the United States, but not in Japan. The mainstream product in Japan is the one and one-half horsepower heat pump.

From 1975 to 1979, not much of significance was done to the product. The sheet metal work improved efficiencies a little but mostly resulted in reduced material costs. The mechanical design at this point was being led by a U.S. company. In 1980, Mitsubishi Electric introduced a product that used integrated circuits to control the air-conditioning cycle. In 1981, one year

EXHIBIT 11
Changes in Features of Mitsubishi Electric's 3 hp Heat Pump

Year	Model Number	Cooling EER (BTU/W.hr)	Added Features or Major Changes in Features
1976	PCH3A	7.4	
1977	PCH3B	7.8	Sheet metal
1979	PCH3C	7.8	Remote control
1980	PCH3D	8.0	IC for control and display
1981	PCH3E	8.0	Microprocessors for 2-wire connection and quick connect freon lines
1982	PCH3F	8.9	Rotary compressor, louvered fin, inner-fin tube
1983	PCH71AD	9.9	Expanded electronic control of cycle
1984	PCH80AD	7.1–11.5	Inverter
1985	NA	7.1–11.5	Shape memory alloys
1986	NA	8–12.5	Optic sensor control
1987	NA	8–12.5	"Personal Pyramid"
1988	NA	8–14	Learning defrost and setbacks

later, Mitsubishi replaced the integrated circuits with micro-processors, resulting in a product that is simple to install and very reliable.

In 1982, Mitsubishi introduced a new version of the product with a high-efficiency rotary compressor in place of the very dated reciprocating compressor. The condensing unit of this new product had louvered fins and inner fin tubes for much better heat transfer. All the electronics were changed because the balance of the system changed. The EER improved markedly. In 1983 Mitsubishi expanded the electronic control of the cycle by adding sensors and more computing power to the unit and markedly improved EER again. In 1984 Mitsubishi introduced a new version of the product with an inverter to enable the motor speed to be controlled, making possible an even higher EER.

By 1990, the features of the Mitsubishi three-horsepower heat pump include logic circuits that enable the machine to "learn" its defrost cycles and perform them automatically. Further, the units can "learn" the user's pattern for adjusting temperature and will mimic this pattern. Also, the unit has an electronic air purification pack.

That is a lot of change in 10 years. It is change accomplished one year at a time. It is change that has carried Mitsubishi and its Japanese competitors who also make similar product improvements to the position of technological leadership in the global residential air-conditioning industry.

In 1985 the management of the U.S. company was debating whether or not to use integrated circuits in its residential heat pump. If the decision had been to do so and the typical four to five years were consumed developing and introducing the new product, the company would introduce the new product in 1989 or 1990. The new product would have been technologically equivalent to the products that the Japanese placed in their markets in 1980. The U.S. company, and its U.S. competitors, are 10 years behind the Japanese in the design of residential air conditioners.

The management of the U.S. company followed the lead of many U.S. companies who had lost the technological and innovative leadership in their industries. They sourced air conditioners and components from their Japanese competitors. There is much talk today of the "hollowing of America." The phrase is a reference to the movement of manufacturing from the United States to lower-wage countries and the former manufacturer's concentration on the remaining sales and distribution functions. But examples such as residential air-conditioning represent the true hollowing of America. This hollowing is the result of lost technological and innovative leadership, supposedly America's long-term competitive advantage.

More companies will be hollowed until they can reduce their new-product development and introduction cycles from 3 to 4 years to 12 to 18 months. This is the requirement to stay in competition. Even faster cycles are required to set the pace of competition—to seize the initiative.

THE STRATEGIC IMPLICATION OF TIME COMPRESSION

The strategic implications of time compression are significant. As time is compressed:

- Productivity improves.
- Prices can be increased.
- Risks are reduced.
- Share increases.

As time is compressed, productivity improves. The productivities of four competitors, three Japanese and one German, in the automobile business are compared in Exhibit 12. Productivity, expressed as output per employee, is related to improving responsiveness. Responsiveness in this industry can be approximated by work-in-process turns. Note that the absolute level of productivity does not mean much because the companies have differing levels of vertical integration. What is meaningful, however, is the consistent pattern in the slope of each line. As each competitor doubles its WIP turns, its productivity improves by 30 to 35 percent. This rate of improvement is realized whether the competitor is Japanese or Western and whether the competitor is number 1, number 2, number 3, or even number 4. The other revealing insight of this chart is that there is an experience effect to improving responsiveness. Mazda is trying to catch Nissan and Nissan is trying to catch Toyota. The company

EXHIBIT 12

As Time Is Compressed, Productivity Increases

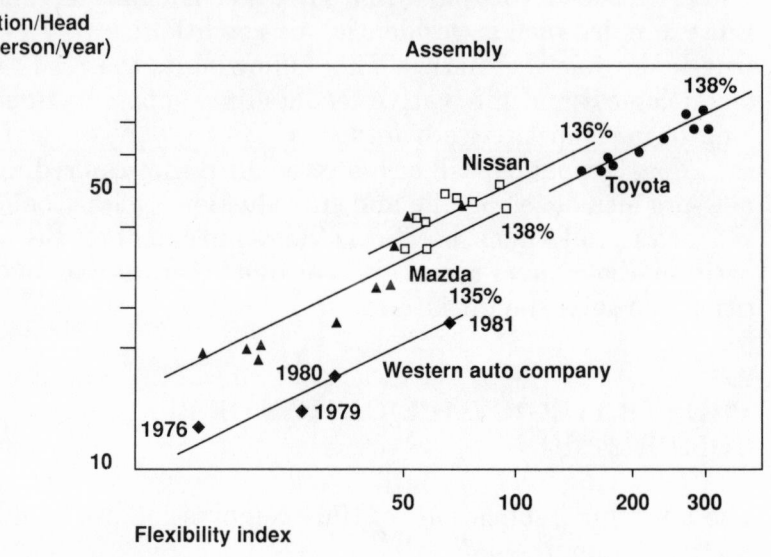

The Boston Consulting Group.

out front has the advantage, as long as it does not sit on the advantage. The company that starts late must play "catch-up" and incur all the costs of becoming responsive, while not obtaining many of the benefits accruing to the leader.

As time is compressed, prices can be increased. Time-based competitors that have a response advantage over their industry colleagues can charge substantial price premiums. In the examples shown in Exhibit 13, the companies that established a response advantage over their competitors (ranging from 2:1 to 9:1) can charge a price premium of 20 to 100 percent. Their customers are willing to pay such a premium not only for the faster service but also because they, in turn, can resell the goods quickly and therefore can make more money because of faster turns despite lower margins that result from higher prices.

When time is compressed, risks are reduced. As has been discussed, the risks of new-product development are greater when a

EXHIBIT 13
As Time Is Compressed, Prices Can Be Increased

	Time Advantage	Price Premium (%)
Electrical components	2:1	59
Locks	3:1	25
Writing papers	5:1	20
Commerical doors	9:1	100

company has a slow product development cycle. There are surprising amounts of risk in all businesses, even beyond product development. Risk costs money but seldom does it show up on the balance sheet and income statement in an obvious or continuous fashion. It shows up usually in a discontinuous fashion in the form of a write-down of some sort. The fashion business is a business fraught with risks. Conceptually, however, the fashion business is a remarkably simple business. Management has to get only two things right to succeed. Management has to make sure that what it is making is selling and that it is not making something that is not selling. Everything else is extraneous. If a company can get those two things right, it can experiment all it wants with look and positioning. If one product fails, the company can abandon it and push the winners.

Much of the apparel sold in North America is sourced from abroad. The buyers of retail chains that source from abroad must place orders for new merchandise 9 to 12 months before they intend to sell the goods. The capability of forecasting nonvolatile demand is greater than for highly volatile demand. Thus, for apparel items that do not have a high-fashion content, such as men's slacks, these long lead times do not present much of a challenge. However, for goods that are very fashionable, such as women's blouses and skirts, demand forecasting is very difficult. Indeed, retail buyers' forecasts are off by as much as 40 percent on average for goods sourced abroad. For goods sourced from typical onshore manufacturers who need four months lead time, the forecasting error falls to 20 percent. For goods sourced from special quick make-and-ship suppliers who can respond within two weeks, the forecasting error is negligible. When all the costs of sourcing abroad are considered, including the costs of over- and underforecasting, the onshore purchase of high-fashion apparel can be readily justified.

As time is compressed, share can be increased. Some successful companies have hit market share ceilings. Customers do not want to buy any more from them. The companies do a good job, but the customers are still worried about the risks of missed deliveries or unresponsiveness. This can be very frustrating for companies because it forces them to either diversify or accept no or low growth. Reducing price to gain more market share when share is already high does not generate healthy profits on the margin for the corporation.

When companies compress time, ways can be found around market share ceilings. One manufacturer of industrial webbing is a good example of a company that has used time-based competition to increase market share. Industrial webbing is the fabric that runs through the inside of machines to carry intermediate products through the equipment. Depending on the size of the machine, this webbing can cost from $1,000 to $250,000 a piece. Because the webbing is usually expensive compared to the cost of the machine it runs on, most shop managers keep one in the machine and one in storage. When the web breaks, another one is pulled out of storage, and a new one is ordered.

This company has the dominant share of the business in the United States and could not get any more of the business. When the customers were asked why they did not give them more share, their answer was that although they have delivery lead times of 22 weeks, they were not reliable. Their delivery lead times were 22 weeks plus or minus 10 weeks. Customers felt that they needed a second source because they could not afford to risk their machines coming down because of a failed web.

The company reduced its response time to eight weeks plus or minus a day. The factory was virtually unchanged. All the significant changes occurred in the sales cycle, the engineering, the scheduling, and the system incentives, goals, and measurements for all the participants in the sales, order entry, engineering, and manufacturing support processes. This company's volume shot up 16 percent, and their share soared to 37 percent.

This year they will close out with a lead time of six weeks, plus or minus a day, their volume will be 41 percent higher, and their share will be up to 45 percent.

TIME-BASED COMPETITION

The high ground of competition today is being taken by companies who are responding faster and increasing variety at a greater rate than their competitors. Those companies not holding the high ground in their markets are slipping into commodity-like competition where price is the principal differentiating competitive variable. Many companies in commodity businesses are being badly battered. Other companies are investing to increase the technological sophistication of their products and escape the low-end competition by "moving upscale." This is a global dynamic.

Japanese companies relying on exports for growth are under enormous pressure to respond to the increasing strengths of other Asian competitors and of the yen. In the automotive industry, for example, Japanese companies are being pressured at the low end of the market by Korean companies and the high yen. The middle market is plagued by overcapacity. The profit outlook is uncertain. The Japanese, as well as Western automotive companies, must have processes that perform well to escape cost competition with the Koreans.

However, the performance of Western and Japanese processes varies considerably. Western companies require 14 to 30 days to build vehicles. The Japanese need two to four days. The sales order and distribution cycle in the United States is 16 to 26 days. The Japanese cycle is typically six to eight days and is as low as four for Toyota. In the United States new-vehicle design and introduction take four to six years, in contrast to the two and one-half to four years needed in Japan. The median age of the product offering is five years in the West and three years in Japan. As a consequence, customers are being offered choices that can have significant technical differences.

For instance, compare the Chevrolet Berretta and the

Mazda 626. The base price of both cars when introduced was about $10,000. However, the level of technology in the Mazda overwhelms that of the Berretta. With the Mazda, one has a choice of three or four valves per cylinder and a choice of turbocharging or even supercharging. The Berretta has the basic, dependable cast iron engine block and head with two valves per cylinder. The Mazda's transmission options include an electronic automatic and four wheel drive. Mazda offers an electronically adjustable suspension and variable assisted power steering, as well as four wheel steering. The Berretta has none of these options.

And the pattern remains unbroken. The new Lexus by Toyota, which was introduced in the United States in late 1989, is technologically equivalent to the BMW 7 series. The Lexus sells for about $35,000; the BMW 7 series sells at a price in excess of $70,000.

As companies, particularly Japanese companies, increase the pace of competition, management must modify its focus from cost and quality to cost, quality, variety, and responsiveness. Faster response is the key. The investments required to significantly reduce the consumption of time in an organization result in lower cost, higher quality, and an increasing ability to accommodate greater variety. Time is the most fundamental variable of business.

To become a time-based competitor, a company must first and foremost reduce the delays in its value-delivery system as far as possible. This is more than a factory task; the factory seldom accounts for more than 40 percent of the total time a customer is forced to wait. The process often must be extended to the sales and distribution system and to the process of developing new products and innovation. The result of reducing delays is improved responsiveness and increased variety.

The ability to increase variety and enhance responsiveness requires new strategies to capture value. Management must first make the value of its new capabilities explicit by segmenting its customers by their sensitivity to responsiveness and pricing accordingly. Time-based companies must let competitors have those customers that are willing to wait a long time for the best price.

Second, a company has to surprise its competitors, and it has to claim the high ground. Surprise means not telling the world about the changes being made. The company will need time to establish a response advantage over competitors. And once it has established this advantage, the company will need time to obtain the most attractive customers and induce them to be dependent on the company. Obtaining commitment from the most demanding customers—those that want what they want when they want it—is achieving the high ground of competition. Finally, the longer competitors can be kept from following, the longer the first flexible competitor will reap the profit advantages of its new responsiveness.

Significantly increasing the responsiveness of an organization by reducing its consumption of time creates opportunities to outmaneuver the competition and attack indirectly. When a company can outmaneuver its competition it can be proactive rather than reactive, and the competitors are pushed into expensive defensive postures. Those executives that seize the opportunity the fastest will reap the greatest rewards.

Time-based competition is a phenomenon of today. While most visible among Japanese companies, fast response is being used as a competitive weapon by Western companies—albeit smaller companies in markets that are less visible than automobiles and electronics. But virtually all businesses can use time as a competitive weapon. Time is an equal opportunity competitive advantage. All competitors have the same amount of time. The one difference among competitors is how productively time is used.

CHAPTER 4

SPEED AND PRODUCTIVITY

Roger W. Schmenner

Editor's Note: Professor Roger Schmenner of the School of Business, Indiana University, has extensive experience in manufacturing strategy. His article with Robert Hayes in the *Harvard Business Review*, "How Should You Organize Manufacturing," has influenced the thinking of many of today's manufacturing executives. In the 1980s, his widely adopted textbook, *Production/ Operations Management: Concept and Situations*, shaped the thinking of thousands of business students about operations.

This chapter by Roger Schmenner is based on a recent, extensive and worldwide survey of manufacturing practices. Some of the details of that study, including some of the statistical findings, are given in tables in the appendix to this chapter.

INTRODUCTION

This study into the strategic advantages of speed was instigated by what seemed to be a perplexing problem. During the early 1980s, America witnessed a steeply declining rate of growth in productivity, but "productivity"—what it meant and how you got it—was confusing. At that time, concern about growth rates prompted a rash of "solutions" to the U.S. productivity decline relative to other industrial countries (see Exhibit 1). In particular, plant managers were bombarded by a host of programs and ideas for improving their plants' productivity performance: those that became known as MRP, CIM, SQC, JIT, and OPT, and other programs of automation, people motivation, industrial en-

EXHIBIT 1
Productivity Comparisons across Countries (Index of Output per
Hour—1977 = 100)

Country	1960	1970	1975	1980	1982	1984	1986
United States	60.0	79.1	93.4	101.7	103.6	116.6	126.0
Canada	50.4	77.0	91.2	102.3	100.3	110.2	112.1
Japan	22.0	61.4	85.3	125.4	127.6	152.2	168.2
France	39.8	70.1	87.9	112.6	122.3	133.8	140.9
West Germany	40.0	68.2	89.0	109.8	114.7	123.5	131.4
Italy	36.5	72.7	91.1	116.9	122.6	134.7	138.4
United Kingdom	55.6	80.5	94.6	108.1	118.2	129.5	138.2

Souce: Bureau of Labor Statistics, U.S. Labor Department.

gineering, value analysis, factory focus, reorganization, and
gain-sharing plans. It seemed as if everybody was touting at
least one of these as *the* way to go to increase factory productivity so as to put America "on top again."

Yet, it was clear that a plant could not institute all of these
programs, certainly not all of them well. Grappling with this
problem led one to confront basic questions: Is any one of the
proffered solutions more fundamental or more important than
the others? What are the true gems among the myriad of programs and initiatives that a factory's management could em-
bark upon? These questions led to the endeavor of trying to
differentiate the best performing factories from those that were
"run of the mill." Just how were factories performing on various
productivity measures? What characteristics of their products,
plants, and programs could explain differences in the performance of factories in diverse industries? Attacking this issue of
productivity entailed collecting and analyzing data from hundreds of factories on their actual experiences. This chapter reports the findings produced by that research effort.

A WORLDWIDE SURVEY OF
MANUFACTURING

The research effort studied plants in all areas of the United
States, Europe, and Korea. To a lesser extent, plants in South
America, Australia, and other parts of Asia were represented.

The U.S. study was conducted from 1984 through 1986. A survey mailed to plant managers collected 265 responses to more than 220 questions. The mailing was purposely designed to oversample plants of large companies, given their importance to industrial America and the evident similarities of many small plants. Half of the questionnaires mailed (about 9,000) concentrated on Fortune 500 companies, while the other half reached randomly selected company plants.

The second portion of the U.S. study involved visits to 26 plants within two loosely defined industries: (1) computer manufacturing, and (2) vehicles and vehicle parts manufacturing (mainly automotive and farm equipment). The two-day visits at each plant consisted of tours and interviews with the plant manager and several other managers. These interviews collected plant information on productivity gains and on all aspects of plant operations over the previous five years, to the extent that the plant kept such data. These plant data permitted the construction of a "total factor" productivity index, perhaps the most general indicator of factory performance.

During the 1986–87 academic year, a slightly modified version of the U.S. questionnaire was sent to an international array of plants selected from an extensive mailing list provided by IMEDE (a premier international business school in Lausanne, Switzerland). Usable surveys were returned from 130 plants located in 30 countries. Most of these surveys came from within Europe, although the remainder came from every corner of the world, including Asia, South America, and Australia. At the same time, Professor Rho Boo Ho of Sogang University (Seoul, South Korea) translated the American questionnaire into Korean and conducted the survey among a sample of South Korean plants. That survey secured 160 usable responses from those plants. Thus, the results reported here reflect phenomena that are truly international in character.

In this research the most important productivity measure was that of labor productivity gain. Labor productivity, often defined for just the direct labor in the factory, is the most widely tracked measure of productivity, and it is very reliable. Only an unusual situation could cause a more general measure of productivity (such as total factor productivity) to consistently demonstrate a trend that is not captured by labor productivity.

The surveys asked for the latest year-to-year gain in labor productivity at the plant, by whatever measure was used. About two-thirds of the factories used a classic measure of output per unit of input, while about one-third used efficiency measures (e.g., actual vs. standard) or some other, nonclassic definition. Although these latter measures of productivity varied enormously in their definition, their correlations with the labor productivity measure were high.

PRODUCTIVITY DIFFERENCES ACROSS PLANTS

Results obtained from statistical analysis indicate why some plants outperform others. These results derive from regression analysis performed on the full sample of plants in the three data sets. Comparative statistics from this detailed analysis are given in Exhibit 3 in the appendix. Two stipulations about the results are needed. First, given the wealth of variables covered in the questionnaire, it was not possible to enter into a regression all of the variables that could possibly affect productivity gain. Second, given that some plants did not complete all of the survey questions, the inclusion of many variables would have caused a serious reduction in the number of observations used in the regression. Thus, the 10 variables reported as "what does matter" are only those that were found to be statistically significant. The information that is not included as significant, however, is every bit as important managerially as that which is statistically significant. These points, therefore, are taken up next.

What Does Not Matter

Certain variables appear *not* to be important to an explanation of labor productivity differences across manufacturing plants. The most notable among these variables are the following:

1. *Age of the Plant.* While the age of the equipment in the plant makes a difference, the age of the plant itself does not. Old plants seem to do just as well in productivity advance as young plants.

2. *Union or Nonunion Labor.* Union status of plants was investigated only for the United States, but the U.S.-specific results confirm that union status, in and of itself, is insignificant in explaining varying gains in production by different plants.

3. *Material Requirements Planning.* Whether a plant operates with a material requirements planning system (MRP) or, indeed, an MRP II system, is not important in explaining productivity differences. This is not to say that planning a plant's material requirements is not an important endeavor. Rather, it simply says that productivity differences apparently do not result from the varying mechanisms used to determine those material requirements. Class A users of MRP, all other things being equal, do no better than Class Z users of MRP.

4. *Wage Incentives.* How workers are paid does not seem to influence productivity gains. Whether the company provides an incentive wage scheme, be it a group incentive plan or a gain-sharing plan, or whether it simply pays by the hour does not seem to matter. One pay plan is about as good as any other for realizing productivity gains.

5. *Sun Belt versus Frost Belt Location.* The results obtained from American plants reveal that, all other things being equal, a Sun Belt location does not impart any advantage for gains in labor productivity. Similarly, findings from the IMEDE data showed that a Northern European location was no better than a Southern European one, and that an Asian location was no better than a European one.

6. *Conventional Industrial Engineering.* Approaches that make use of time and motion studies do not seem to have a significant impact on labor productivity gain. Indeed, labor productivity gain actually suffers to the extent that the plant relies on "staff" for productivity-enhancing ideas. The "staff" variable has three components: top management, mid-level management, and industrial engineering. In fact, the effects of industrial engineering, considered by itself, reach statistical significance on the negative side. This result indicates that, all other things being equal, reliance on industrial engineering for the primary ideas for making productivity gains is actually disadvantageous for the factory. Labor productivity gain appears to be a broader concept.

What Does Matter

What are the trends and themes that carry over from America to Europe to Korea? Based on the survey results, what variables tend to explain differences in productivity?

1. *Throughput-Time Reduction.* One of the questions asked in the survey was: For the major product line, how long is the cycle time for production (the throughput time) from release of an order to the factory floor until the shipment or warehousing of that order? A follow-up question asked: How has this production cycle time changed over the past five years? How much shorter, or longer? The results clearly state that the greater the reduction in throughput time, the greater the labor productivity gain for the factory. This result rings true for the entire sample of 555 plants and also for each of the three data bases separately.

2. *New Technology.* The importance of investment in new technology is captured by two distinct variables: the average age of the factory's equipment and the degree of the factory's involvement in process hardware advance. The first variable was estimated using the plants' responses to a question which asked for the percentage of the plant's equipment that was in each of five separate age categories: 0–2 years, 3–5 years, 6–10 years, 11–20 years, and more than 20 years. The second variable was derived from responses to the following question: Characterize this plant's involvement in process hardware advance: (1) heavy, entire sections of the plant devoted to new technology; (2) modest, some new technology in operation; (3) tinkering, occasional trial machines in operation; or (4) few tangible advances made to date. The results show that the younger the average age of plant equipment and the more advanced the plant in process hardware, the greater the plant's labor productivity gain, all other things being equal. This technology factor is particularly evident for the Korean plants, as one might expect for a developing country like Korea. It is less of an influence in both the United States and Europe.

3. *Work Force Participation and Simpler Plant Organization.* The survey asked the factory to "Label with an 'A' the *chief* areas of recent productivity accomplishment. Label with a 'T' those areas that are key targets for future productivity gains."

One of the 22 items that could have been so labeled was "Increased worker participation in management decisions, say, through worker teams." One result obtained from these items is that factories with increased worker participation tend to have greater productivity gains. Complementing that result are two results that argue for fewer labor grades in the work force and for fewer levels of management in the plant. Both of these results stress simpler plant organization.

4. *Less Inventory.* The lower the plant's level of inventory, the greater its productivity gains. This result comes from examining the change in the inventory turns of the factory during the past five years. The greater the increase in turns, the greater the increase in productivity.

5. *Improved Quality.* The importance of product quality comes out in the perceived strength of the factory's quality control/assurance function vis-à-vis other functions. A survey question asked the factory to identify its three strongest functions and its two weakest functions from a list of 14. The stronger the perception of the plant's quality function, the greater the productivity gain achieved.

6. *Improved Purchasing.* The plant's purchasing function demonstrates its importance in the same way as the quality function. The stronger the perception of the purchasing function, the better the productivity gain.

7. *Competition on Product Performance.* Those plants that perceived themselves as competing against others on the dimension of product performance apparently gained more productivity than did plants that perceived themselves as competing on other dimensions such as cost, delivery, new-product introduction, customization, or flexibility.

8. *Different Products Competing Differently.* One question in the survey asked whether the major product lines of the factory competed in different ways, that is, on cost, delivery, product performance, and so on. If, in fact, the products did compete in different ways, the plant's productivity performance proved to be better. Exactly why this is so is not obvious, but may be an indication of the plant management's understanding that different skills are required by different product lines.

9. *Plant Staff not a Good Source of Productivity Ideas.* As mentioned earlier, to the extent that the plant counts primarily on its staff for productivity-enhancing ideas, the worse plant gains appear to be.

10. *Overtime.* The more overtime hours worked per production worker per week, on average, the higher the labor productivity gain. This result may be an indication of high-capacity utilization, which, by many measures, increases productivity. This variable was particularly evident in the data from Korean plants.

The pattern of these 10 results prompts one to ask anew: "Which one of many possible ways (cited in the introduction) to increase productivity is most important? Is it CIM? Is it factory focus? Is it MRP? The one way best supported by these results is "Just-in-Time manufacturing." Almost all of the results are consistent with the tenets of JIT. Nowhere in the three surveys was JIT mentioned, yet the aspects of JIT appear in force in the conclusions which can be drawn from the survey findings.

LESSONS FROM THIS RESEARCH—THE SPEED OF THE TORTOISE

Given this research, the conclusion is that the horse for manufacturers to ride is the one in the colors of JIT and, more specifically, the banner of throughput-time reduction. Many manufacturing experts are now coming to the view that the best short definition of JIT is the contraction of time in all aspects of business operations—for example, throughput times, new-product introduction times. The value of contracted throughput time for increased productivity is an incontestable result of this research.

A good analogy here is the "tortoise versus the hare" (see Exhibit 2). Aesop had it right; he'd make an excellent plant manager. The tortoise and the hare, in this context, is more than a fable: It has some real applications to modern factory management. The hare's factory would have fast equipment, but not necessarily well maintained. (How many companies take pride in the fact that their equipment runs faster than the rated capacity?) Materials wait to be worked on in the hare's factory.

EXHIBIT 2

The Hare's Factory

Fast equipment, but not well maintained

Materials wait a lot to be worked on

Lots of WIP   Lengthy setups

Scattered layout; materials move a lot, often
into and out of the storeroom

Jerky stop and start schedules, often interruped

No help sought of others

The Tortoise's Factory

Slow equipment, always able to run

Materials do not wait long to be worked

Quick setups   Little WIP

Compact, rational layout; material moves
economically between operations

Smooth schedules, few interruptions

Help asked of all

The hare simply does not get off its haunches all the time to run. That means a large amount of work-in-process inventory, lengthy setups, and a scattered layout where materials move around a lot, in and out of the storeroom and all over the factory. Production schedules, when actually followed, are jerky, start-and-stop affairs, similar to the way in which the hare runs. Schedules are interrupted often to do something else that has a momentarily higher priority. And, because the hare has no voice, no help is sought from others inside or outside the factory.

What about the tortoise's factory? How does the tortoise win the race? The tortoise's factory is filled with conventional equipment that is slow but always able to run. Materials do not wait to be worked on; once they are in motion, they always remain in motion, just like the tortoise ran its race. Work-in-process inventories are held down and lengthy setups which can interrupt the flow of the product through the factory are reduced. The tortoise does not reduce throughput time by running faster machines; rather, it reduces throughput time by making sure that the product never stops. Machines must operate only fast enough to meet the pace of demand itself; anything faster than that does not provide an advantage. The tortoise's factory also avoids the stop-and-start production found in the hare's factory. Rather, the tortoise's factory is characterized by compact, rational layouts where materials move economically—just like the tortoise did—by taking the shortest trail between two points. Production schedules are smooth, with few interruptions. Tortoises may not be able to speak, but the "factory tortoise" seeks help from all involved, including suppliers. Everybody becomes involved in helping the tortoise find his way through the maze of possible trails to the finish line.

It is possible to be more specific about the benefits of throughput time reduction as a banner for the factory. The benefits are both direct and indirect in nature. The direct benefits are those to which the pursuit of reduced throughput times inevitably lead. Pursue throughput-time reduction and certain actions must result. Among these benefits are the following.

Better Quality
Given a proper accounting of time, the reduction of throughput time helps force a factory into better quality. Doing it right the first time saves time. If one measures throughput time properly, it should be a weighted average of the times required to make all of the units in a batch of the product. Such a weighted average includes not only those units that finish quickly without having to be reworked or scrapped, but also those units of the initial batch that spend time in the rework area. And, if some units are scrapped and that causes another order to be cut to complete what was planned, then the "catch-up time" should be counted.

Low Inventories

Keeping inventories low is a key way to remove time within the factory itself. Materials frequently spend inordinate, and unnecessary, time sitting in inventory.

A Rational, Balanced Process without Bottlenecks

Management wants a process that is in balance, with as much waste as possible removed and with layouts that are very compact. Bottlenecks are never desirable. Nor are most excuses for bottlenecks viable, like this often-used one: "Well, that bottleneck is caused by the mix of products that we put through the factory this month." A factory must work to make the process flexible to product mixes so that bottlenecks simply do not occur.

Diminished Chaos and Confusion in Scheduling

Expediting is a sin. Manufacturers must eschew expediting because that practice interrupts throughput time. Trickling out a seemingly endless stream of engineering change orders for a product is another way to add confusion to the factory and to increase throughput times.

These four important direct benefits flow from reducing throughput time in the factory. Equally as important as these direct benefits, though, are three indirect benefits.

Elimination of Overhead

Most companies are presently struggling to control overhead as their key cost factor. Overhead rates are skyrocketing. Many remedies to high overhead costs have surfaced, among them computer packages and the reorganization of overhead functions. These approaches can help, but cost efficiency actually comes from removing the need for overhead. How is this done? The manufacturer must make the process as "clean" as possible with as short a throughput time as is feasible. If materials travel a clean, swift route through the factory, the factory can start to shed overhead. When materials flow swiftly through the factory, the manager can consider eliminating some of the production control, inventory control, and inspections along the way. Information will not be entered into the computer system as many times and products will not be inventoried as often.

An 80/20 rule seems to apply in this regard: 80 percent of the time and effort spent by the factory's overhead is devoted to just 20 percent of the products that are made by the factory. That 20 percent is likely to be composed of products that linger in the factory, meandering from station to stockroom to station and back. However, if plant managers are provided incentives based on throughput-time reduction, those managers will work on those products that take long times to process. Their evaluations will depend on time reduction, and they will find clever ways of improving the manufacturing processes of even the low runners of the factory.

Indeed, the managers may lobby for a change in the cost accounting system, because changing the cost accounting system can provide even more motivation. Why not abandon the usual practice of allocating overhead to products based on direct labor and, instead, allocate it on the basis of throughput time? The more quickly the product goes through the factory, the less overhead (per dollar of product value) it attracts. The more slowly a product moves through the factory, the more overhead it attracts. Managers who want to be evaluated favorably would then pay attention to collapsing throughput times, and that is completely beneficial. When managers start focusing on time, the factory can start shedding overhead that is now only allocated, not systematically removed.

Quick Response to the Marketplace
The market shares of a large number of companies would jump dramatically if their order-to-market times could be cut substantially. Quick delivery often is highly valued in the marketplace.

Improved Capital Appropriations
When capital appropriations are requested, the sharp pencils come out and trimming begins. The traditional economics for these decisions should be augmented with another measure: throughput time. If an investment in new equipment is made,

will throughput time come down or not? If throughput time may not be reduced by the investment in new technology, the firm should think twice about that technology. Islands of automation are not what is needed because islands are hard to get to and away from. Making time an element of the capital appropriations process can help keep new technology an integral part of the process, and thus much more advantageous to the company.

In summary, speed—researched here as factory throughput time—is strategically advantageous. Statistically, speed can be shown to be an important component of productivity gains, and speed—the speed of a tortoise—has the added advantage of tying together a host of related benefits (e.g., quality, low inventories, diminished chaos) in ways that personnel on all levels in a manufacturing organization can find persuasive. Increasing American productivity need no longer be perplexing; the "speed of the tortoise" can help win the productivity race.

APPENDIX: DESCRIPTION OF WORLDWIDE SURVEY OF MANUFACTURING

The research effort encompassed plants in all areas of the United States, Europe, and Korea. To a lesser extent, plants in South America, Australia, and other parts of Asia were represented. Expansion of the study beyond America, except for Korea, was financed by IMEDE (a premier English-language business school in Lausanne, Switzerland, with extensive international connections). Survey data for Korean plants were collected by Professor Rho Boo Ho of Sogang University in Seoul, South Korea.

The U.S. study was conducted from 1984 through 1986. It was the most elaborate because it involved the construction of two distinct data bases: one that used a mail survey of plants of all kinds and sizes, and one that required visits to collect extensive five-year histories from plants in two selected industries. The questionnaire mailed to plant managers collected 265 usable responses to more than 220 questions. The mailing was purposely designed to oversample plants of large companies, given their importance to industrial America and the evident similarities of many small plants. Half of the questionnaires mailed (about 9,000) concentrated on Fortune 500 companies, while the other half reached randomly selected company plants. Returned surveys constituted just 3 percent of those mailed out, but no biases occurred in either the size or location of the plants from which surveys were received.

The second portion of the U.S. study involved visits to 26 plants within two loosely defined industries: (1) computer manufacturing (essentially factories that stuffed printed circuit boards and put them into something that either is marketed as a computer or is so close to being a computer that it could be called one), and (2) vehicles and vehicle parts manufacturing (mainly automotive and farm equipment). The two-day visits at each plant consisted of tours and interviews with the plant manager and several other managers. These interviews solicited ideas about productivity advance and collected plant information on productivity gains and on all aspects of plant operations over the previous five years, to the extent that the plant kept such data.

These plant data permitted the construction of a "total factor" productivity index, perhaps the most general indicator of factory performance. Fortunately, the results obtained from the use of that particular productivity measure parallel closely the results obtained from the survey data and other aspects of the study. This outcome is important because it lends credence to the validity of the European and Korean studies that relied entirely on mail survey data without plant visits.

During the 1986–87 academic year, with the cooperation of IMEDE, a slightly modified version of the U.S. questionnaire was sent to an international array of plants selected from IMEDE's extensive mailing list. Usable surveys were returned from 130 plants located in 30 countries. Most of these surveys came from within Europe, although the remainder came from every corner of the world, including Asia, South America, and Australia. At the same time, Professor Rho Boo Ho in Seoul translated the American questionnaire into Korean and conducted the survey among a sample of South Korean plants. That survey secured 160 usable responses from those plants. Thus, the results reported here reflect phenomena that are truly international in character.

The most important of the productivity measures examined was that of labor productivity gain. Labor productivity, often defined for just the direct labor in the factory, is the most widely tracked measure of productivity, and thus is important in its own right. Although labor productivity is a biased measure, it is very reliable. Only an unusual situation could cause a more general measure of productivity (such as total factor productivity) to consistently demonstrate a trend that is not captured by labor productivity.

The surveys asked for the latest year-to-year gain in labor productivity at the plant, by whatever measure was used. About two-thirds of the factories used a classic measure of output per unit of input, while about one-third used efficiency measures (e.g., actual versus standard) or some other, nonclassic definition. These latter (more general) measures of productivity varied enormously in their definitions; however, their correlations with the labor productivity measure were high. Because different plants define labor productivity in different ways, actual levels of labor productivity as reported could not be compared. Only the percentage gains in productivity from one time period to another could be compared.

Statistical analysis of the three data sets collected from the U.S., IMEDE, and Korean plants revealed many differences in the plants of major producers among the world regions represented by the survey samples. Comparative statistics for selected plant variables examined are detailed in Exhibit 3.

American manufacturing, compared to both Europe and Korea, uses smaller factories that are controlled as profit centers, fewer levels of management, and less labor turnover, but significantly more labor grades. The production plans of the American plants, however, are more likely to be in tumult—that is, more fluctuation in the schedule and fewer days when the schedule is frozen. More than the others, U.S. plants reported competing on quick delivery and receiving orders directly from customers. American labor cost as a percentage of sales is comparatively high, although throughput time is notably low. Equipment age,

EXHIBIT 3
Comparison of Plants: American, IMEDE, and Korean Sample Means

Plant Characteristic	American Sample	IMEDE Sample	Korean Sample
Employment	457*,†	1020*	1498†
Year plant was started	1959.3*,†	1953.9*,‡	1969.8†,‡
Estimated average age of equipment	11.6 yrs.*,†	8.3 yrs.*	7.6 yrs.†
Number of product lines	13.2	15.1	10.3
Percent accounted for by major line	63.4%*,†	50.8%*	52.0%†
Percent indicating products made elsewhere within company	56.5%	46.5%	50.3%
Percent feeding plant's output to other plants within company	36.6%	41.1%	37.3%
Average number of part numbers	6,013	14,936	6,392
Percent of process that is line flow or continuous flow in character	52.7%†	46.4%‡	59.7%†,‡
Percent of plants competing largely:			
On cost	30.6%†	33.3%	40.4%†
On quick delivery	50.6%*,†	28.7%*,‡	14.3%†,‡
On flexibility	25.3%*,†	35.7%*,‡	16.1%†,‡
On customization	40.4%†	33.3%‡	67.1%†,‡
Percent of plants receiving orders directly	71.9%*,†	50.4%*,‡	36.0%†,‡
Typical month-to-month fluctuation in schedule, in percent terms	28.3%*,†	19.5%*,‡	10.1%†,‡
Days during which schedule is frozen	12.1*	20.6*	17.2
Typical number of engineering change orders in a month	22.2*,†	40.9*,‡	10.2†,‡
Typical rework percentage	5.1%	4.2%	4.3%
Typical throughput time, in days	29.1†	43.1	62.6†
Percent evaluated as a profit center (rather than cost center)	68.3%*,†	31.8%*	41.6%†
Average levels of management at plant	3.6*,†	4.0*,‡	4.9†,‡
Average number of labor grades	10.0*,†	5.6*,‡	4.0†,‡
Labor turnover	6.1%†	7.9%‡	19.5%†,‡
Labor cost as percentage of output value	19.5%*,†	15.4%*	13.8%†
Average inventory turns	12.8*	8.4*,‡	16.3‡
Average gain in labor productivity	7.0%†	7.5%‡	11.7%†,‡
Average gain in a more general measure of productivity	9.1%†	8.0%‡	15.0%†,‡

* - a statistically significant difference at least at the 5% level for the American and IMEDE samples.
† - a statistically significant difference at least at the 5% level for the American and Korean samples.
‡ - a statistically significant difference at least at the 5% level for the IMEDE and Korean samples.

although not factory age, is the highest of the three groups surveyed. In general, these characteristics of the "typical" American plant seem to ring true with much of the conventional wisdom about U.S. manufacturing.

As expected, the European plant is the oldest in the three samples and is significantly larger than the American plant. In contrast to the others, however, the European plant is least likely to be a profit center and most likely to suffer from problems associated with supplying diverse products to fragmented markets. Comparatively, the European plant's production process is less of a "flow process," its inventory turns are the lowest, its level of enginering change orders is the highest, and its perceived need to compete on flexibility is the greatest.

Compared to plant averages in the other two samples, the Korean plant is by far the largest and the newest, and it uses the newest equipment. These characteristics are in keeping with Korea's position as a rapidly developing country dominated by a select group of major conglomerate manufacturers. While throughput time is high in Korean plants, inventory turns also are high. Product lines are fewer, and competition is more on cost and customization. Orders are less likely to be received directly at the plant. Schedules and engineering changes are less likely to be disruptive. While labor costs as a percentage of sales are low, labor turnover is high and labor grades are few. Korean plants, on average, reported significantly greater productivity gains than did plants that comprised the American and IMEDE samples.

PART 2

NEW-PRODUCT DEVELOPMENT

CHAPTER 5

NEW-PRODUCT DEVELOPMENT: THE NEW TIME WARS

Joseph D. Blackburn

Speed in manufacturing must be coupled with rapid deployment of new products to stay abreast of the competition. In the global market, speed in manufacturing cannot sustain a competitive advantage once the product reaches the twilight of its life cycle. Heated competition hastens obsolescence, and the edge in the marketplace shifts to the firm with the new product that has the technology and features that consumers crave.

Who are the speed merchants in new-product development (NPD)? Barry Bebb, vice president for systems architecture at Xerox, claims that most of the leading firms are Japanese. He states that the Japanese dominate U.S. competitors on two key dimensions: (1) high-quality, low-cost, timely products through their Just-in-Time (JIT) manufacturing systems; and (2) the rapid commercialization of technology in the form of new products.[1] To investigate this further, a group of manufacturing executives recently was asked to perform a simple experiment: make a list of the world-class leaders in NPD and, then, make a second list of the fastest manufacturers, those that have mas-

---

[1]H. Barry Bebb, "Quality Design Engineering: The Missing Link to U.S. Competitiveness," Keynote Address to the NSF Engineering Design Research Conference, Amherst, MA, June 11, 1989.

tered JIT. To the surprise of no one, most of the firms on both lists were Japanese. A sprinkling of North American firms were on both lists—firms such as Hewlett-Packard, Motorola, Northern Telecom, and GE appeared frequently; however, the predominant names were Honda, Toyota, Sony, and Canon. The startling observation was that the two lists of most of the executives were virtually identical! In their view, the firms with time-compressed manufacturing were also the world's fastest new-product developers, thus extending Bebb's hypothesis that the same Japanese firms dominate the twin dimensions of manufacturing and product development.

The thesis of this chapter is that the high correlation observed between speed in manufacturing and NPD is not coincidental. The skills needed to win the race are the same in both cases. Firms that have successfully converted conventional manufacturing systems to JIT have mastered the managerial arts and the process sciences that also serve to time-compress their cycles for new-product development. These two processes have more than speed in common. The steps that a manufacturing operation must take to move to JIT are remarkably similar to the path that should be followed to achieve rapid new-product deployment.

This chapter examines the NPD process closely and explores the links to JIT. Unlike JIT, research on NPD is in the embryonic stage. Just-in-Time has been studied extensively, and there is a set of generally accepted operating practices that should be followed to attain world-class, short-cycle manufacturing. However, there is currently no comparable set of guidelines for NPD, although some concepts are beginning to crystallize. Fortunately, one can extract from the research on JIT and the small amount of research on product development to synthesize some guidelines. Case studies of successful new-product introductions are underway at several academic centers;[2] Kim Clark and his colleagues at Harvard have studied the development

---

[2]Manufacturing research centers at Boston University, Indiana University, University of Michigan, University of Minnesota, and Vanderbilt University all have industry studies in progress.

of new automotive products extensively; the National Science Foundation is funding several interdisciplinary studies of the design process. The early evidence from these endeavors suggests that the fast new-product developers are the "learning organizations"[3] that have cultivated their skills in manufacturing and are now propagating them into other functions of the firm, particularly new-product design. As with JIT, firms striving to achieve faster product introductions must adopt and emulate the techniques of the leading-edge firms—the same firms that lead the way in revolutionizing manufacturing.

A key characteristic of the time-based competitor is the potential to overwhelm adversaries with an accelerating stream of innovative products or services. The market potential of this phase of time competition is examined in the next section. Subsequently, the new-product introduction process is dissected and examined. The linkages to JIT are established and sharp comparisons are drawn between conventional American practices and those of Japanese competitors. The concluding section develops a set of guidelines based on JIT principles for accelerating the speed of new-product introductions.

NEW PRODUCTS: THE ESSENTIAL COMPETITIVE WEAPON

"Better never than late!" For new-product development, the old saying must be inverted to have meaning in today's markets. Late to market means lost profits: profits foregone during the part of the product cycle in which the firm can least afford to lose them. Premium prices can be charged in the early stages of the new product's life cycle before competitors have brought their technology on-stream. Significantly, the largest profit margins accrue under the umbrella created by being the only producer in the market with the newest product.

A 1987 study in the automotive sector by Clark, Chew, and

---

[3]See, for example, Robert Hayes, Steven C. Wheelwright, and Kim B. Clark, *Dynamic Manufacturing: Creating the Learning Organization* (New York: Free Press, 1988).

Fujimoto indicates that, for the case of a $10,000 car (a small one by today's standards), each day of delay in introducing a new car model into the market represents, conservatively, $1 million in lost profits.[4] Another study by Clark[5] indicates that the Japanese can, on average, complete a new-car development project about 18 months faster than either their U.S. or Western European competitors. So, if a U.S. firm and a Japanese firm each started on a new-car project at the same time, then, with average performance, coming to market a year and a half behind the Japanese firm could cost the lagging U.S. firm on the order of a half billion dollars in lost profits: an invisible killer that, as an opportunity loss, does not appear on the profit and loss statement.

In the United States, the new-product wars have moved to higher, more expensive ground: the luxury car market. In 1989, Toyota and Nissan weighed in with new automobile lines—the Lexus from Toyota and the Infiniti from Nissan. At the same time, Honda moved to steal the scene by introducing the *second* generation of its Acura line. So Toyota and Nissan, making their maiden entry into a market already crowded by established West German, British, and U.S. competition, found that Honda suddenly slipped a new, stacked deck into the game.

The new Acura is another example of how Honda uses rapid new-product deployment as a competitive weapon, a weapon that has been sharpened through years of honing while Honda cut a swath through the motorcycle industry.[6] In the automobile industry, product life cycles are historically six to seven years; Honda, on the other hand, revamps its line every three or four years, and the big wheel is turning faster. Honda's speed is important to its dealers, who love it. According to one dealer, "They

---

[4]Kim B. Clark, B. Chew, and T. Fujimoto, "Product Development in the World Auto Industry," Brookings Papers on Economic Activity, 3, 1987, pp. 729–71.

[5]Kim B. Clark, "Project Scope and Project Performance: The Effect of Parts Strategy and Supplier Involvement on Product Development," *Management Science*, 35, no. 10 (October 1989), pp. 1247–63.

[6]See James C. Abegglen and George Stalk, Jr., *Kaisha: The Japanese Corporation* (New York: Basic Books, 1985), pp. 46–52, for an account of the Honda-Yamaha motorcycle wars in which Honda assailed Yamaha with new products.

[Honda] can turn products around and respond to a changing market much faster."[7] The strategy pays off: Honda entered the luxury market with the Acura in 1986 and, by 1988, Acura sales had exceeded the combined sales of Mercedes, Jaguar, and Porsche.

Getting to market first is more important for profitability and sustainable advantage than staying within the development budget. Recognizing this, John Young, the CEO of Hewlett-Packard, recently challenged his managers to cut the time between product conception and breakeven in half.[8] Using a model developed by consultants, Stephen Hamilton of Hewlett-Packard reports in Chapter 8 that, in terms of profitability, being six months late on a project is much worse than being 50 percent over budget.

Late to market can, in fact, be fatal. The global video-cassette recorder (VCR) market is an excellent case in point. The first widely available, affordable video cassette recorder was introduced in 1972 by Philips, the Dutch electronics conglomerate. Japanese firms did not become significant players in the market until the mid-1970s. By the end of the decade, Philips was no longer the market leader and, today, its VCR has vanished from the market. When Philips introduced its second-generation VCR near the end of the 1970s, Japanese competitors such as Panasonic, Sony, and Hitachi had already *introduced and retired* three generations of VCRs. Philips was blasted out of the market by a ponderous new-product development process.[9]

Rapid new-product development also makes market research much easier, particularly in fashion businesses, in which products have short shelf lives. The trick is no trick at all: Offer the customer a glimpse at a menu of new products, note what they like, then rush it into the market. Sony was able to leverage new-product development speed into a deadly marketing weapon in the infant stages of the compact disc (CD) market by

---

[7]Joseph B. White, "Honda, Trying to Outpace Its Rivals, Unveils New Acura Luxury Cars Today," *The Wall Street Journal*, May 4, 1989.

[8]Brian Dumaine, "How Managers Can Succeed Through Speed," *Fortune*, February 13, 1989.

[9]Bro Uttal, "Speeding New Ideas to Market," *Fortune*, March 2, 1987.

offering an array of disc players with a variety of features at different price points. They quickly observed a surge in demand for low-price, mass-market versions and had the product development speed to exploit this expanding market segment. The ability to respond to market shifts with new products is really the secret to Benetton's success in the trendiest market of all: fashions for teen-age girls.

In the rapidly expanding market for electronic pagers ("beepers") and mobile phones, Motorola has shown that U.S. manufacturers can use "first-to-market" capability to turn the tables on Japanese competition. Motorola designed a new pager as well as the factory to produce it in 18 months, half the usual 3 years.[10] This swiftness enabled the firm to regain a position as market leader in the pager market. With its new Micro-Tac phone, Motorola extended this marketing capability to the mobile phone market. Because the Micro-Tac is 33 percent lighter than the next lightest mobile phone from Matsushita, Motorola has been able to charge a premium price of $2,500 and is still not fulfilling all the demand.[11]

The swift will continue to grab the profits in markets created by the emergence of new electronic products. Although high-definition television (HDTV) is still in the research and development stage, the battle lines are being drawn. Fearing a recurrence of the recent experiences with DRAM chips and other electronic products, in which slow development by U.S. firms caused them to flee from a scene dominated by Asian competitors, many U.S. industrialists and politicians are lobbying for governmental subsidy of a manufacturing effort in quest of HDTV ("If we can put a man on the Moon, we can . . ."). Even with massive support, a U.S. effort in HDTV is likely to be too little and too late. Japanese firms already have a huge lead in developing HDTV technology and that lead should widen during the time required for U.S. firms to move from product concept to production. Japanese electronics firms are traditionally among the world's fastest at bringing an existing technology to the

[10]Brian Dumaine, *Fortune*, February 13, 1989.
[11]Lois Therrein, "The Rival Japan Respects," *Business Week*, November 13, 1989.

marketplace. Coupling that speed with their running start in R&D on HDTV, Japanese electronics firms should own the market by the time a U.S. product enters.

A CLOSER LOOK AT THE NEW-PRODUCT PROCESS

Documenting a firm's new-product development process is an order of magnitude more difficult than documenting its manufacturing system. To see why, first construct a process flow diagram or "road map" for the manufacture of a specific product and then construct a similar flow diagram for the design and development of some new product. In manufacturing, components can be tracked from machine to machine and each step in the operation can be observed and charted: There is a "thing" to be followed, something tangible to be seen, measured, tested, and so forth. In new-product development there is no "thing" to be tracked. The development process primarily involves the transformation of information, often in electronic form or intangible mental images. Charting a diagram for a process as nebulous as new-product development, therefore, can be a daunting task. Nevertheless, this step is necessary if progress is to be made toward compressing time in this vitally important function.

The development process varies across firms and industries but, despite this diversity, there is general agreement on the major stages of the new-product development process. These stages, as shown in Exhibit 1, reflect findings from research carried out by Clark and Fujimoto.[12] In their view, new-product development is an information-processing exercise that has four basic stages: concept generation, product planning and design, product and process engineering, and pilot tests and ramp up to production.

Although U.S. firms and Japanese firms follow this se-

---

[12]Kim B. Clark and Takahiro Fujimoto, "Overlapping Problem Solving in Product Development," in *Managing International Manufacturing*, K. Ferdows (ed.) (Amsterdam: North-Holland, 1989a).

EXHIBIT 1
Stages of New-Product Development

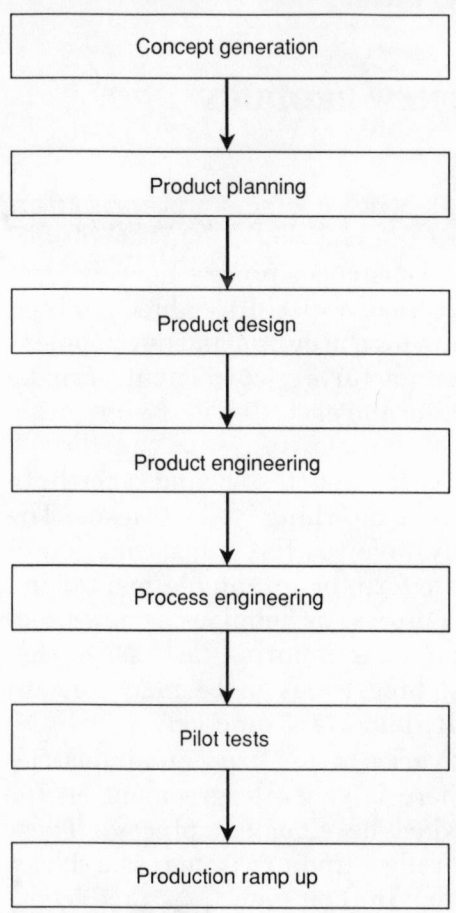

```
┌─────────────────────────────────┐
│      Concept generation         │
└─────────────────────────────────┘
                │
                ▼
┌─────────────────────────────────┐
│       Product planning          │
└─────────────────────────────────┘
                │
                ▼
┌─────────────────────────────────┐
│        Product design           │
└─────────────────────────────────┘
                │
                ▼
┌─────────────────────────────────┐
│      Product engineering        │
└─────────────────────────────────┘
                │
                ▼
┌─────────────────────────────────┐
│      Process engineering        │
└─────────────────────────────────┘
                │
                ▼
┌─────────────────────────────────┐
│          Pilot tests            │
└─────────────────────────────────┘
                │
                ▼
┌─────────────────────────────────┐
│      Production ramp up         │
└─────────────────────────────────┘
```

quence of stages in roughly the same manner, they do not complete them at the same speed. Exhibit 9 in Chapter 3 presents the results of a comparison by George Stalk of the Boston Consulting Group of the efforts made by two firms, one U.S. and one Japanese, to develop a new marine gear. Although the technology being introduced was similar, the Japanese firm was faster at every stage except the design review process. This outcome tends to confirm the common observation about the Japanese style of management: Managers tend to devote a lot of time in the decision process to make sure that everyone is on board and agrees with the decision; once this consensus is reached, everyone rows together, and new-product projects move more quickly than those of their Western counterparts.

Steps such as detail design and prototype production, Stalk found, were up to four times slower at the U.S. firm. This does not mean, however, that designers in Japan are four times faster or that they can build prototypes that much faster. The answers lie deeper than differences in the speed of engineers. Typically, the U.S. design team must produce more designs than a Japanese team before obtaining an approved design. More designs are needed because of more engineering changes (the manufacturing equivalent of rework); thus, the excessive design time is related to the quality of design information. Delays in prototype production also can be traced to engineering changes and other problems deeply rooted in manufacturing. A specific, in-depth look at prototype problems confronting U.S. firms is provided by a case study of an air compressor manufacturer that follows this marine gear discussion.

In terms of total time for the marine gear product (Exhibit 9, Chapter 3), observe that the Japanese project was finished in about half the time taken by the U.S. project. Assuming that the U.S. firm had made the technological advances needed to begin the project at about the same time as the Japanese firm, the latter would have been in the market with the new product a full year and a half ahead. Eighteen months in the market provides opportunities to charge a price premium, to garner higher profits, and so forth. When the U.S. firm enters the market with its marine gear product, it may discover, to its dismay, that the Japanese competitor can respond by cutting prices because the 18 months of lead time has allowed the firm to trim the fat from its manufacturing process.

More ominous is the fact that by replicating the speed of its first product development project, the Japanese firm (emulating Honda's feat in the luxury car market) could have completed a *second* project while the U.S. competitor was developing its first product. After 18 months of being alone in the market with new technology, the Japanese firm, conceivably, could bring a second new product to market to coincide with the first U.S. product introduction and, thereby, render the U.S. entry obsolete the day it hits the market.

A Case Study in Prototype Production

Campbell-Hausfeld, a major U.S. producer of air compressors, faced a problem that plagues many firms, namely, long delays in the time required to produce and test prototypes. In designing a new air compressor for the consumer market, the firm found that the longest delay was in the production of a prototype. Speed was essential because a competitor was nearing market entry with a similar new product, and the possibility loomed that Campbell-Hausfeld would be the last to market.

To short-circuit that time, the firm considered two options: prototype production in the United States versus prototyping offshore in Hong Kong. Exhibit 2 compares the project plans under these two alternatives. Even considering the delays in shipping designs across the Pacific, coordination with subcontractors in Hong Kong, and the like, the offshore alternative was projected to be four to five months faster. Most of the time problem resided in the tool and die stage to obtain plastic injection-molded parts and die-cast components. According to the project manager, "The longest lead times were in plastic injection-molded parts. We're located in the heart of the greatest concentration of injection-molding firms in the United States, and they all quote four- to five-month lead times. On top of that, they don't seem to care whether they get our business or not. Now, when you talk lead times in Hong Kong, they say, 'When do you want it?' The service you get is almost like having a suit made. They are ready to work round-the-clock to get your business."[13]

Despite the time disadvantage, management still leaned toward having the work done locally in order to stay on top of engineering changes and other potential problems in the tooling process. When the tooling costs were totaled, however, the decision clearly swung in favor of offshore procurement. Tooling

---

[13]Interview with project manager by Mike Siebenmorgen, July 1988.

EXHIBIT 2
Prototype Development: United States versus Hong Kong

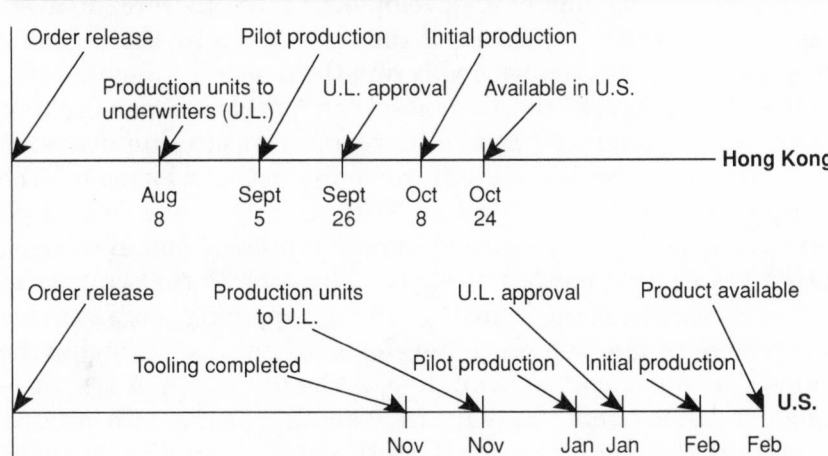

costs in the United States were more than double the cost quoted in Hong Kong: a difference of $220 thousand.

The Hong Kong option was twice as fast at half the cost. This is what other researchers report from comparisons of new-product development efforts between Japanese and American firms. Campbell-Hausfeld's experience tends to support those observations. This case strongly suggests that some of the problems that U.S. firms have in new-product development are linked to weaknesses in manufacturing. It also helps explain why many firms are desperately seeking ways to eliminate the dependence on prototypes.

Data from other industry studies tell the same story: U.S. firms lag far behind in their ability to speed new products to market. A major study of new-product development in the world auto industry carried out by Kim Clark of the Harvard Business School makes this point clearly.[14] In this study, Clark compares

---

[14]Kim B. Clark, "Project Scope and Project Performance: The Effect of Parts Strategy and Supplier Involvement on Product Development," *Management Science* 35, no. 10 (October 1989), pp. 1247–63.

the strategies of U.S., Japanese, and European automakers in 29 automotive development projects and evaluates the results of each project. The Japanese development projects were, on average, completed in two-thirds of the time taken by the U.S. and European projects and with only one-third as many engineering hours. On average, the Japanese can develop a new vehicle about 18 months faster than either Americans or Europeans.

Why does the West lag in new-product development? The story now unfolding is familiar. With JIT, Japanese firms first developed, and then unleashed, time-compressed manufacturing processes as a competitive weapon. Those firms that cultivated JIT processes in manufacturing are now applying those skills to the processes of new-product development and are achieving the same outcomes: speed, flexibility, and high quality. The new-product deployment systems in leading Japanese firms are, therefore, the result of technology transfer within the organization. Management skills sharpened in manufacturing have flowed by diffusion to the product planning function.

To examine in detail why the West lags, conventional Western practices in both new-product development and manufacturing were compared with their Japanese counterparts. The results of this comparison are striking and insightful: Western practices in new-product development and manufacturing bear a strong resemblance to, but differ sharply in significant ways from, their Japanese counterparts. The typical U.S. new-product development process seems to have been cut from the same cloth as conventional batch manufacturing. The key features of new-product development that have been emerging from Japan are virtually identical to those of JIT manufacturing. Given that these processes were created by the same managers ("from the people who brought you large-lot manufacturing"), it is not surprising that the conventional Western practices in the two functions evolved along similar paths. Both U.S. systems were synthesized by a generation of managers that focused on centralized administrative control and built formal, departmentalized organizational structures and accounting systems to pursue an objective of cost minimization.

Side-by-side comparison of manufacturing and new-product development yields needed insights into how U.S. product development must be changed. First, a comparison is drawn between conventional Western practices in manufacturing and new-product development. Then the similarities between practices for these two functions that have emerged from Japan are examined. These two sets of comparisons reveal a clear path that U.S. firms must follow to achieve world-class speed in new-product development.

CONVENTIONAL MANUFACTURING AND NEW-PRODUCT DEVELOPMENT IN AMERICA

Exhibit 3 summarizes the similarities between conventional practices typically found in manufacturing and new-product development in U.S. firms. Each component of these practices is subsequently discussed.

Batch Sizes

As discussed in Chapter 2, long changeover times are the economic driver for the large batch sizes that pervade U.S. manufacturing. The long lead times found in conventional manufacturing are causally linked to the time lags and work-in-process (WIP) inventory created by large-lot production. Slow changeovers also drastically limit the flexibility of the production system to respond to shifts in market demand.

New-product development, as Eastman has pointed out, is essentially a "problem solving activity within the framework of an information processing model."[15] In studies of new-product design in the United States, researchers have found that, as in manufacturing, information is commonly processed in large batches—that is, a "phased approach" to problem solving.[16] The analog of the changeover in new-product design is the change from working on one design problem to another. In conventional

---

[15]C. Eastman, "On the Analysis of Intuitive Design Processes," *Emerging Methods in Environmental Design and Planning*, G. T. Moore (ed.). (Cambridge, Mass.: MIT Press, 1972).

[16]Kim B. Clark, B. Chew, and T. Fujimoto, "Product Development in the World Auto Industry."

EXHIBIT 3
Comparison of Western Practices in Manufacturing and New-Product Development

Process Parameters	Conventional Batch Manufacturing	Conventional (Western) New-Product Development
Batch sizes	Large batch production; transferred in large batches	Information processed and transferred in large batches
Layouts	Job shop	Functional by department
Process flow	Sequential activities	Sequential
Scheduling	Centralized scheduling	Centralized control
Employee involvement	Low	Low
Supply relations	Little coordination with suppliers	Low involvement with suppliers in design of components
Quality	High defect ratios; rework	Numerous engineering changes; rework
Automation	Islands of automation; isolated robots or transfer devices	Isolated systems (e.g., CAD systems with negligible intergration)
Lead times	Long	Long

design, an engineer or designer tends to work on a large chunk of the problem, reach a resolution, and then send it on to the next function. In a study of the use of CAD/CAM in product design, Paul Adler found that "drawings often sat idle, waiting to be worked on, just as long in electronic form as in paper form."[17] In conventional batch manufacturing, value-added time usually constitutes less than 5 percent of the total production time; over 95 percent of the time, the product is either sitting in batches awaiting processing or being moved to another work center to get "in queue" for more processing. Adler's findings suggest that Western approaches to product design have characteristics dis-

[17]Paul S. Adler, "CAD/CAM: Managerial Challenges and Research Issues," *IEEE Transactions on Engineering Management* 36, no. 3 (August 1989), pp. 202–15.

turbingly similar to large-batch manufacturing. Consequently, throughput time in the development shop and adaptability to changing design requirements are retarded by designs-in-process (DIP) inventory.

Layouts and Process Flow

In conventional batch manufacturing the job shop layout, in which machines are grouped by function, prevails. Machining centers are spaced far apart to accommodate the enormous piles of WIP inventory. In new-product development, the layout is analogous to the job shop: Groups tend to be located in offices according to their function. Product engineers occupy one office, process engineers reside in another, and designers are located elsewhere; each is grouped by activity or skill instead of collectively by project.

In both cases the partitioned, spatial layout encourages sequential performance of development activities. A large batch of work is finished and then transferred to the next department where it waits in a queue for further processing. Product design is an "over the wall" process. For example, the design department completes its work on a project and then tosses the design over the wall to engineering. Communication between work groups is inhibited by the layout.

In a study of U.S. and Japanese die-making operations in the auto industry, Clark and Fujimoto found that the U.S. die production process was organized like a "job shop"[18]—that is, in a process or functional layout. Dies worked their way around the shop in a haphazard flow that depended on the processing requirements of the specific die. Each work center completed all of its work before transporting the die to the next center. The study pointed out that die making lies on the critical path and, thus, influences the ultimate completion time of the project. Significantly, Clark and Fujimoto's data show that U.S. firms tend to take twice as long for the die-making process as do Japanese firms.

---

[18]Kim B. Clark and Takahiro Fujimoto, "Overlapping Problem Solving in Product Development."

Information flows in conventional new-product development are like a one-way street: Information flows in one direction only, downstream from designers to product and process engineers and from engineering to manufacturing. There is little
feedback from downstream. No information from downstream
about the "quality" of design information is transmitted back
upstream.

Management and Employee Involvement

Conventional manufacturing offers no incentives for shop floor
employee involvement in planning and control functions. Schedules come from "above." Process control is centralized, characterized by computerized, MRP systems and shop floor scheduling algorithms. With large-batch production, teamwork or
coordination between work centers is effectively discouraged.

In the United States, new-product management also tends
to be centrally controlled and scheduled. Because functions are
compartmentalized, the personnel assigned to the project team
are dispersed and rarely are in direct communication. This functional detachment often is a formidable obstacle to teamwork
and coordination: designers design, engineers engineer, and
marketing sells—top management guides the process, so there
is no need for communication between the groups. This approach
subverts the teamwork that enhances speed in product development.

Project management techniques, such as PERT and CPM,
embody the way most Western executives think about managing large-scale projects such as new-product development. The
success of these methods by defense contractors in "crashing"
the time to develop new weapons systems, such as the Polaris
missile, is now part of the lore of project management. The fundamental idea of these techniques is to control project completion time by focusing on the "critical path" of activities: that
sequence of tasks whose completion determines the total project
duration.

The project management paradigm, however, contains several traps that can hinder their application to new-product development. First, the critical-path diagram, per se, induces a
mindset in which sequential activities prevail. As part of the

planning process, an arrow diagram of the project is constructed that depicts (at minimum) the activities, estimated activity times, and the precedence relationships that exist among activities. Once committed to paper or chart, the diagram clearly shows that task B (which might be design review by manufacturing) can begin as soon as task A (design) is completed. The snare is that, by implication, task B *cannot* begin before task A is completed. Conceptually, the diagram becomes an obstacle to achieving overlapping activities and to transferring information in small batches: the very type of design process that Clark and others claim is much faster. In other words, CPM and PERT diagrams reinforce a managerial mindset that views new-product development as a sequential process.

The CPM and PERT diagrams used as a conceptual tool for new-product development have a second flaw: an emphasis on centralized, rather than local, control of the project. Local control by the project team fosters the speedy communication of ideas and information among team members. In centralized project management, managerial control is exercised from the top down, and the project diagram bolsters that form of management by centralizing information in the hands of a project manager.

Supply Relations

Traditional supplier relationships in the United States tend to be at arms length and, at times, adversarial rather than cooperative. In a comparative study of U.S. and Japanese supply chains, Welch, Blackburn, and Jueptner found that U.S. manufacturers were less likely to form long-term, cooperative relationships with their parts suppliers than were their Japanese counterparts.[19] In the United States, suppliers tended to be farther away and, consequently, made less frequent and larger shipments of components. Inspections were more likely to be required for incoming shipments, and manufacturers tended to

---

[19]James Welch, Peter Jueptner, and Joseph D. Blackburn, "Buyer/Supplier Relations: A Contrast between Japan and the United States," Operations Roundtable Working Paper, Owen Graduate School of Management, Vanderbilt University, 1989.

rely less on single-sourcing of components. Relationships were cemented with contracts, rather than trust. Consequently, the level of communication and information sharing between manufacturer and supplier was much lower in the United States than in Japan.

Traditionally, Western manufacturers rarely involve their suppliers in new-product development. The relationship is characterized by the same lack of mutual cooperation and communication as is found in the supply of components. Suppliers first see a component after it has been designed, when they are asked to bid on a supply contract. No attempt is made to take advantage of the supplier's technology or design expertise, and that constitutes waste of a valuable resource for time-compressing the development cycle.

Quality

Both conventional batch manufacturing and conventional product design are beset by quality problems for reasons traceable to sequential, large-lot production. In batch manufacturing, components are produced in bin loads which are then hauled to the next station. Defective parts, obscured by the size of the batch, are simply passed on to the next function. The view on the shop floor is: "What's the problem with a few defective components in a large lot? They can correct it later. It's not my problem." This myopic attitude, as most of the gurus of quality have pointed out, inhibits the awareness of quality needed on the shop floor if defects and rework are to be eliminated. Moreover, large batches also increase the time between occurrence of the quality problem and detection. Quick detection of defects is necessary to determine the cause of the quality problem. Pinpointing problems at the source allows for corrective action and promotes overall quality improvement by preventing a recurrence of the problem.

In product design, quality means getting it right the first time. Not getting it right means rework (or reworked designs)—a major cause of long lead times in development cycles. Batch and sequential processing of information in the design process amplifies the effect of rework on the project lead time because it extends the time between problem inception and detection. The situation at France Division of Scott-Fetzer, a manufacturer of transformers and appliance timer devices, as described by Tom Caste, director of engineering, typifies the problem: "Marketing is involved with designers in the early stages of the project in setting new-product specifications. Then marketing doesn't see the design again until the designer is finished and a prototype has been built. Often marketing discovers that design compromises have been made that reduce the market appeal of the product. If marketing is really upset, then we go back for another design iteration, and that can take months."[20] Similar problems can occur if process engineers deal with manufacturability issues sequentially, after the design is complete. Firms frequently discover that components, designed separately, cannot be easily assembled, and this also results in additional design iterations. The sequential process of design, review and inspect, and redesign to correct errors of misspecification or misunderstanding adds months, even years, to the new-product development cycle.

Automation

Automation in conventional U.S. manufacturing has been applied sporadically and inconsistently. Many managers, seeking to ward off the threat of offshore competition, seized on automation as a way to leapfrog the competition with technology. Often, "islands of automation" were created when a robot was inserted, like a surgical implant, into the production layout to relieve some bottleneck or deal with a quality problem. No effort was made to prepare the process for automation; in many cases, firms simply succeeded in "hardwiring" an inefficient, obsolete process.[21]

In product design, similarly, automation has been applied to isolated segments of the process: CAD systems, for example, are

---

[20]Tom Caste, "Getting Started in Time-Compressing New Product Development," Presentation to Time-Based Competition Conference, Vanderbilt University, December 1988.

[21]Robert A. Millen, Joseph D. Blackburn, and Edward Popper, "Is the Company of Today Ready for the Factory of the Future," Owen Graduate School of Management Working Paper No. 86–36, Vanderbilt University, 1986.

widely used to automate the design function and in plant re-layout tasks. Throwing a work station into the lab without making changes in the organizational structure or the design team carries the same risk as tossing a robot onto the factory floor: It may result only in adding expensive technology to a process that may need to be revamped or discarded. In his studies, Adler found that "in most cases, CAD/CAM procedures had merely automated the paper system without any redesign of the actual information flow path or even any change in the allowed processing times. The need to reconfigure engineering procedures to improve . . . engineering operations efficiency is urgent in most of the sample businesses."[22]

COMPARING JUST-IN-TIME MANUFACTURING AND NEW-PRODUCT DEVELOPMENT IN JAPAN

To complement the analysis of Western practices, JIT manufacturing practices were compared with the emerging Japanese procedures for new-product development. Exhibit 4 summarizes this comparison. As will be discussed, the emerging Japanese procedures for new-product development are remarkably similar to JIT manufacturing processes. Both processes are designed for speed, simplicity, and an absolute minimum of waste.

Batch Sizes
Manufacturing by JIT is launched by small-lot production, made possible by reduced setup, or changeover, times. Lot sizes of one are the goal. As the firm refines the production process and moves, asymptotically, toward that goal, it finds that product quality improves, inventory diminishes, and, significantly, lead time shrinks.

In product design, the batch size is measured by the amount of information to be processed or the scope of the problem to be

---

[22]Paul S. Adler, "CAD/CAM: Managerial Challenges and Research Issues," *IEEE Transactions on Engineering Management*, 36, no. 3 (August 1989), pp. 202–15.

EXHIBIT 4

Comparison of Japanese Practices in Manufacturing and New-Product Development

Process Parameters	JIT Manufacturing	Japanese New-Product Development
Batch sizes	Small-batch production; transferred in small lots	Information processed and transferred in small batches
Layouts	Product layout	Grouped by project team
Process flow	Coordinated activities	Overlapping activities; simultaneous engineering
Management	Localized control; high level of employee involvement	Local control; management by project team
Supply relations	Close coordination with suppliers	High involvement with suppliers in design of components; technology and information exchange
Quality	Low defect rates	Few engineering changes
Automation	Integrated systems; automation follows process simplification	Automation of information flows; integrated CAD/CAE/CAM
Lead times	Short	Short

solved. Clark's research on product development in the auto industry suggests that Japanese firms tend to bite off small chunks of the design problem at successive stages; that is, they reduce the information batch size.[23] Clark also specifically examined the relationship between the scope of the project and the time to completion. Scope, or project complexity, is determined

---

[23]Kim B. Clark, *Management Science*, October 1989.

by the number of unique parts and the amount of supplier involvement. Clark argues that the Japanese take specific managerial actions to narrow the effective scope of their new-product projects, and this partially explains their success in obtaining shorter lead times and fewer manhours.

In another study, Clark and Fujimoto draw an important distinction between the "phased" and "overlapping" approach to problem solving.[24] They share Eastman's view that the major activity in product development is problem solving and claim that speed in new-product development depends on the organization's execution of problem-solving activities. In their research they compare the phased and overlapping approaches in terms of their efficacy in compressing a new-product development cycle. The chief differences between the two approaches are shown in Exhibit 5. In the phased approach, common to U.S. new-product development, information about an aspect of product design is passed along only after the batch of work is complete at the prior stage. The procedure resembles the way material in conventional manufacturing is transferred in large lots from one work station to the next.

As the phased approach is to conventional manufacturing, the overlapping approach (used by leading Japanese firms) is to JIT. In the overlapping approach, information is released in incremental units—the batch size for information transfer from one design stage to the next is reduced to the smallest unit possible. As in JIT, the goal is to achieve the smallest possible lot size. The downstream stage begins design work before all work from predecessor stages is complete. To do this, the transition, or setup time, must be reduced because each team member must learn to pick up the work whenever new units of information are transferred instead of waiting for the entire batch to arrive.

Performance of overlapping activities requires that the new-product system be rapidly adaptable to changes. To accomplish this, as Clark points out, new-product development processes must have some of the same capabilities as the manufac-

---

[24]Kim B. Clark and Takahiro Fujimoto, "Overlapping Problem Solving in Product Development."

EXHIBIT 5
Phased versus Overlapping Approach in New-Product Development

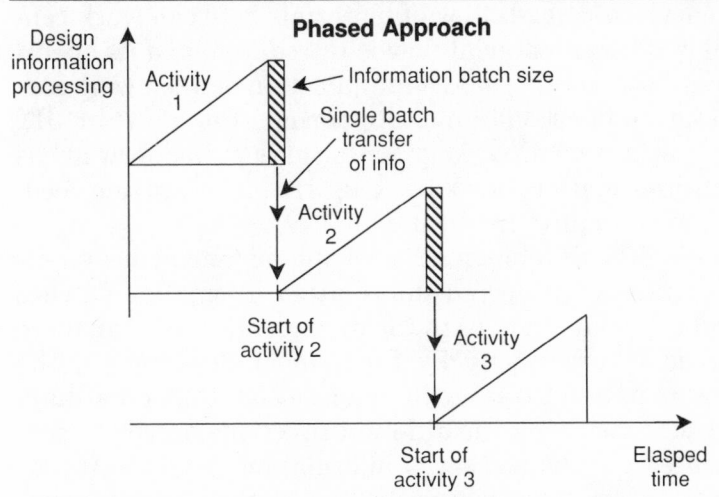

Phased Approach

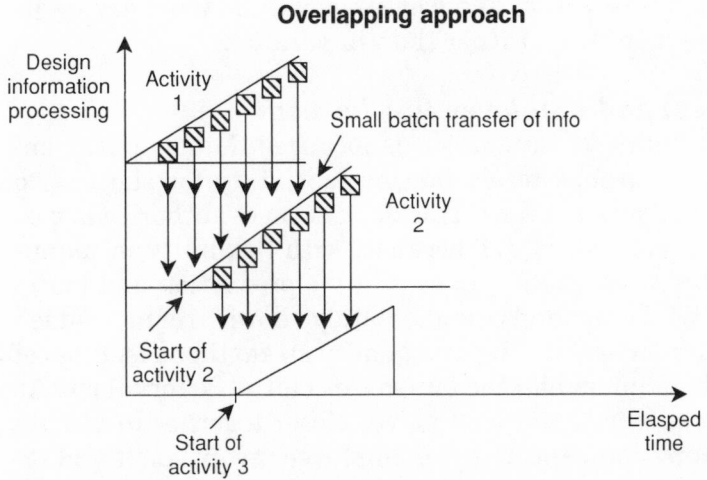

Overlapping approach

turing processes in JIT production systems: multifunctional workers, high quality, local control, reduced setup times, and small batch sizes.

Layout and Process Flow
In JIT, small-batch production forces changes in layouts, material flows, and communication between work stations. Instead of machines grouped according to function, as in job shops, work

stations are arranged for the precise sequence of operations needed for specific products. Compact process flow layouts are formed to ensure a smooth flow of materials between work centers. A pull system of material flow is introduced, and its operation requires close, continuous communication beween work stations. Unlike conventional manufacturing, the flow in JIT operations goes in two directions: Material is pulled forward as needed, but information flows back upstream to provide feedback on material requirements and quality.

In new-product development, overlapping among design activities and transfer of design information in smaller batches fosters similar layout changes in the more progressive Japanese firms. Instead of organizing by function or department, the project team is pulled together in one location, thus creating a "project layout." Bringing the different functions together catalyzes project teamwork and flow of information. A project layout fosters frequent, two-way communication between team members, and this interaction helps break down the tendency to attack tasks sequentially, rather than in parallel.

Management and Employee Involvement

In JIT, scheduling of production is localized. Management encourages work groups to assume responsibility for short-term scheduling and product flow. This delegation of authority is possible, even preferred, in JIT because, with flexibility in manufacturing, the work group can coordinate production and material flow with demands immediately downstream. Added responsibility for scheduling and quality strengthens feelings of team membership among the employees on the factory floor. As space is compressed, everyone moves closer together in a more interactive environment. Greater employee involvement and responsibility for the process promotes a cycle of quality improvement: Employees provide direct, immediate feedback on quality problems and are encouraged to suggest process improvement.[25]

In new-product development, a similar comparison can be

---

[25]A blueprint for developing a "commitment model" in the workplace is provided in Richard E. Walton, "From Control to Commitment in the Workplace," *Harvard Business Review*, March-April 1985, pp. 77–84.

made. In Japan, scheduling responsibility resides with the new-product development team. That is, scheduling responsibility is pushed down to the lowest level. In the United States, by contrast, new-product management tends to be scheduled at the top; coordination is carried out at a high administrative level. In the United States, designers tend to be given specific tasks to perform and to have a limited view of the total project and, consequently, a limited sense of responsibility for the total project.[26] With a team approach, on the other hand, designers and engineers move closer together into a more interactive environment. They have a shared vision of the overall project goal and shared responsibility for the project's success. Instead of working at stations as individuals, workers become part of a team.

Clark's studies of new-product development in the auto industry also highlighted the advantage of coordination and teamwork—both key parts of JIT and, likewise, important in new-product development. This advantage appears in comparisons of the number of new parts or new components that are introduced in a new-product project. By giving additional responsibility to the new-product team, the Japanese firms achieve a high level of coordination that unblocks the inhibitors to information sharing—not only with suppliers but also with engineering teams and other functions within the firm. The resultant knowledge enhances the use of common parts, part kits used in prior models, and so forth. Firms in the United States, conversely, tend to "reinvent the wheel" in each project, particularly when one engineering team is unaware that the desired component exists from a prior project. With team responsibility for product design, design quality tends to improve. Frequent interchange of design information among members of the project team serves as an early warning system for design problems.

Supply Relations
The use of JIT demands supplier involvement not only in frequent deliveries, but also in obtaining cooperative solutions to quality problems. In their studies of supply chains in Japan,

---

[26]Alternative organizational approaches to design and development are provided in James W. Dean, Jr., and Gerald I. Susman, "Organizing for Manufacturable Design," *Harvard Business Review*, January-February 1989, pp. 28–36.

Welch, Jueptner, and Blackburn found that Japanese manufacturers viewed suppliers as partners, rather than adversaries.[27] As partners, they shared information on schedules, technical problems, and new-products concepts. A problem for a manufacturer or a supplier also became the partner's problem; joint responsibility for problem solving was an integral part of the supply-chain relationship.

Japanese companies extend that relationship to new-product development by fostering extensive supplier involvement in the design and engineering of new components. According to Clark, "Supplier involvement (and strong supplier relationships) accounts for about one-third of the manhours advantage and contributes four to five months of the lead-time advantage [held by Japanese automakers]."[28] This suggests synergy in the relationship between JIT and new-product development: The strong supplier relationships that support the shortened lead times in manufacturing also provide a powerful advantage for shortening lead times in product development. A firm's partnership with the members of the supply chain increases the technology and problem-solving resources available to the product development team.

Quality
During most JIT implementations, quality, as measured by the proportion of defective parts produced and the amount of rework, improves steadily as setup times and batch sizes are reduced. As discussed in Chapter 2, the actions taken to reduce setup times simultaneously attack the causes of defective parts. Moreover, JIT promotes a heightened quality awareness at all levels of the organization. Quality problems at any stage become a widespread concern because they can disrupt the entire process. Consequently, problem solving and information sharing becomes a cooperative, team effort. Reaching the goal of synchronized flow in JIT depends on eliminating quality problems.

In new-product development, high quality means minimal

---

[27]James Welch, Peter Jueptner, and Joseph D. Blackburn, Working Paper, 1989.
[28]Kim B. Clark, *Management Science*, October 1989.

rework or redesign. Design quality problems are rarely due to sloppy, incompetent work by designers. Instead, they typically arise because of isolation of the designer and incomplete information or misunderstandings on the part of the designer concerning product requirements. For instance, customer requirements may have been inadequately communicated by marketing (or they have actually changed). Problems in manufacturability may not surface until process engineers attempt to assemble the part with components designed by others. From a quality standpoint, the overlapping approach to design helps prevent rework by transmitting design information bit by bit, instead of waiting until the entire activity is completed. With a two-way information flow between all members of the project team, problems tend to be detected earlier and solved before they contaminate the rest of the project. Designers are integrated into the project team, and the team monitors the progress of the design for adherence to specifications rather than reviewing only completed designs.

Automation
The proper role of automation in a JIT system can be summed up in the phrase "First simplify, then automate." Shigeo Shingo, who helped develop some of Japan's first JIT systems, claims that automation should be used in setup-time reduction *only* after the process has been studied and simplified to remove all the extraneous steps.[29] Automation must not be viewed as a substitute, or shortcut, to process improvement. It is to be employed after a specific process has been smoothed out. A process that has undergone JIT, for instance, is poised to take advantage of automation. Simplifying the process first makes it far easier to automate; robots can make simple movements quickly and unerringly. Automation also can promote system integration through the collection and transmission of information between team members and work groups.

Japanese firms use automation in new-product develop-

---

[29]Shigeo Shingo, *A Revolution in Manufacturing: The SMED System* (Cambridge, Mass.: Productivity, Inc., 1985).

ment similarly to the way it is used in JIT. They recognize that automation alone is relatively powerless to turn around a slow, ineffective product development team. CAD systems can speed up parts of the design function, but if the project is managed in a sequential, departmentalized way, the advantages of automation are likely to be dissipated by frequent engineering changes and redesigns. In the United States, CAD is often viewed as a tool that *encourages* frequent design revisions. The firms that can truly exploit the potential of automation are the ones that first simplify the process of product development. That is, they first build a smoothly functioning project team based on overlapping activities and open, cross-functional communication lines. The power of CAD systems to produce high-quality designs is multiplied in this setting because the evolving design is transmitted electronically through networks and shared with other team members. With CAD/CAE/CAM systems, moreover, it is technologically feasible to automate much of the design, engineering, and machine control programs for manufacturing. The firms best positioned to integrate these computer design activities and accelerate product development cycles with automation, however, are the firms that have already developed a smoothly functioning design team with overlapping activities.

MANAGING NEW-PRODUCT DEVELOPMENT WITH THE JUST-IN-TIME MODEL

In Japan and the United States, firms tend to manage new-product development the same way that they manage manufacturing. The foregoing comparison of U.S. and Japanese management approaches showed a remarkable degree of consistency across the two functions. The traditional Western way of managing new-product development is sequential batch processing; the emerging Japanese model is JIT. The empirical evidence on performance weighs heavily in favor of the Japanese model for new-product development. The leading Japanese firms—Honda, Sony, Toyota, Canon—are the world's fastest at introducing new products. Since JIT yields faster throughput times in manufacturing, it is logical to hypothesize that JIT principles applied to new-product introduction also increase speed in that process.

The JIT model is appealing for several reasons. First, the basic objectives of JIT—eliminating waste in the organization, simplicity, total quality, and speed—are desirable attributes for all of the firm's processes. Second, dedication to the JIT model promotes organizational learning. The lessons learned in manufacturing are transferable to new-product development; knowledge acquired through JIT implementation is recyclable for use throughout the firm. Third, JIT embodies the principle of *Kaizen*, or incremental, but continuous, improvement. Firms that adopt the philosophy of Kaizen tend to attack new-product development in small, manageable chunks—singles, instead of home runs. On the other hand, U.S. firms typically tackle large projects for new products—a radical revision of the product instead of a succession of product enhancements. They go for the home run and, while swinging for the fences, often strike out.

The JIT-based model for the new-product introduction process embraces the key ingredients for successful new-product deployments described in research literature. For instance, the JIT model encompasses elements of Clark's research: overlapping problem-solving activities, project scope, and cooperative efforts with suppliers. The JIT philosophy establishes an environment in which "simultaneous engineering" and design automation can be exploited. In addition, the JIT model demands adherence to the principles of teamwork. The model advocated here, however, is not, and cannot be, a precise, step-by-step blueprint for success. Much remains to be learned about applying JIT in manufacturing, and research on this issue in new-product development has barely scratched the surface. What the present JIT model does, however, is capture within a single framework many of the concepts expressed in the research on record.

The model outlined for managing new-product development is built upon three underlying principles of JIT.

Principle 1: To reduce throughput time, process and transfer smaller batch sizes.

Principle 2: To process smaller batches efficiently, reduce changeover times.

Principle 3: Solve the problems exposed by applying principles 1 and 2.

Inclusion of "Principle 3" is an admission that the model does not offer a "closed form solution" to the new-product development problem. Satisfying principles 1 and 2 unleashes a raft of other problems—quality, layout, information and material flow, and other areas. These problems arise at a similar stage in the evolution of JIT as inventory is reduced and throughput rates increase, and no firms have solved all of them completely. The pursuit of speed, like excellence, is unending.

Principle 1: Reduced Batches of Information

New-product development is essentially an information-processing function in which, as Eastman points out, problem solving is the fundamental activity.[30] In new-product development, the analog of the manufacturing lot size is the batch size of information. In manufacturing, batches of product are made and transported between work centers. In development, packets of information are processed (i.e., decisions made) and transmitted between groups.

Two contrasting approaches to problem solving in product development are phased activities versus overlapping activities. Phased activity is the sequential batch approach to problem solving to which the derisive phrase "over-the-wall" is applied. Phased activities are best suited to functional, departmentalized organizations. On the other hand, overlapping activities are based on processing and transferring information in small batches, similar to small-lot production in JIT manufacturing.

What are the implications of small-batch information processing for product development? First of all, there must be the early release of information from a product designer to a process engineer. Instead of waiting until a final design is complete, partial information is sent downstream so that other members of the project team can begin their work. An obvious advantage

---

[30]C. Eastman, *Emerging Methods in Environmental Design and Planning.*

of this activity is design quality: early detection of problems. With advance information on the design, process engineers can detect problems with materials, manufacturability, or reliability. Feedback on potential design problems promotes early correction before the rest of the design is contaminated. Even more serious problems can be avoided by early transmission of information to marketing and manufacturing. Many of the delays in product development projects can be traced to rework or redesign caused by a misunderstanding of product requirements. Constraints on the project are imposed by all functions. A two-way flow of information is essential to the successful performance of overlapping activities because it communicates the existence of constraints on the product specifications among all members of the department team.

Conversely, serious risks accompany small-batch information processing. Releasing partial information increases the likelihood that erroneous information is sent downstream to other members of the project team. Designers, faced with this uncertainty, must overcome a natural reluctance to release unfinished work to critical eyes. Process engineers, working with incomplete information, must be more flexible and forgiving than in a sequential batch process. Consequently, for this design process to work, continuous communication between team members and high levels of trust must be achieved. Everyone on the project team must understand the new ground rules and be willing to operate in a risky, fluid environment.

Overlapping activities shorten the project cycle time in two important ways. First, parallel activities are stimulated. With early release of information, engineers can begin working on different phases of the problem while final designs are evolving. Not only does this make simultaneous engineering feasible, but also it encourages simultaneous design-engineering-testing activities. The project team works less like a relay team—passing the baton after each lap to the next member of the team—and more like a pit crew at an auto race: everyone working simultaneously toward a single objective. Second, time-consuming rework is avoided because the early release of information promotes early detection that the product design is veering off-target. Mid-course corrections in the design, although lengthening time in the original design phase, tend to yield shorter total elapsed times. The alternative of reviewing completed designs (and finding them off-target) usually leads to several complete iterations of the design cycle and longer elapsed times.

Principle 2: Changeover Time Reduction

Changeover, or setup, time reduction is an apt metaphor for the new-product development process: A new-product introduction is essentially a changeover from an old product to a new one, and the goal is to make the transition as quickly and efficiently as possible. More than a metaphor, the process of reducing a changeover time is an excellent analog for the set of management activities that must be accomplished in new-product development to accommodate the overlapping activities that Clark and Fujimoto espouse. The four basic activities (adapted from Shingo and discussed in Chapter 2) that apply to changeover time reduction in new-product development:

1. Internal versus external activities.
2. Eliminate and simplify tasks.
3. Teamwork.
4. Automation.

Internal versus External Activities
In a setup, speed is gained by eliminating machine downtime. This is done by analyzing every activity to determine if it is internal (in-line) or external (off-line). Activities external to the setup should be done off-line so as not to waste production time. In-line activities should be time-compressed through parallel, or simultaneous, operations and simplification to minimize time loss. New-product development can benefit from the same approach. Overlapping activities help remove much of the work from the critical path; engineers begin working on phases of the problem while design is still being fine-tuned. Much of the work on manufacturing layout and equipment procurement can be converted, with planning and analysis, into an off-line activity. Waiting until all the *i*s are dotted and *t*s are crossed on a prior activity before beginning a critical task extends the completion

time of the project. For example, the approach that GE's Medical Systems Group takes is typical of how many firms are promoting simultaneous activity to reduce lead times: Preliminary design of the manufacturing processes and tests for manufacturability begin at about the same time as prototype construction.[31] Ebert, Slusher, and Ragsdell report that "The simultaneous design (and engineering) process is believed to create more mistakes but leads to fewer iterations of the design steps."[32] Fewer iterations in the design cycle are the reason why simultaneous design and engineering tends to compress project completion time.

Eliminate and Simplify Tasks
In their study of setup times, Shingo and his colleagues found that a large proportion of the time wasted was consumed by machine settings, adjustments, and trial runs of components to get adjustments right.[33] They learned that simple analysis and engineering changes make most machine adjustments superfluous. In some cases, automation was needed, but frequently the solution was achieved by elementary mechanical changes—removing fine-tuning knobs for adjustments and replacing them with simple, physical stops at the desired setting.

In new-product development, astute firms are discovering that many of the delays in design, prototyping, and testing stages can be reduced by eliminating certain activities, such as prototyping, and by simplifying the design. One of the simplest, cost-effective techniques in design is to use off-the-shelf components. Many advanced computer manufacturers, such as Cray and DEC, have designed their new machines to work with standard, off-the-shelf parts.[34] The use of proven components eliminates design time and any uncertainty about manufacturability.

---

[31]Nhora Cortes-Comerer, "Organizing the Design Team," Part III of "Optimizing the Design," *IEEE Spectrum*, May 1987, pp. 41–46.

[32]Ronald J. Ebert, E. Allen Slusher, and Kenneth M. Ragsdell, "Information Flows in Product Engineering Design Productivity," Working Paper, College of Engineering, University of Missouri-Columbia, June 1986.

[33]Shigeo Shingo, *The SMED System.*

[34]Paul Wallich, "How and When to Make Tradeoffs," Part II of "Optimizing the Design," *IEEE Spectrum*, May 1987, pp. 33–39.

Traditionally, the greatest obstacle to the use of standard components has been the designers and engineers themselves, who consider change to be the most important word in their job descriptions.

Can prototypes be eliminated? This question is an important one and merits special consideration at this point. Many firms today are asking the question: Are prototypes passé? In the case of the air compressor prototype described earlier, building that prototype added six months and several hundred thousands of dollars in cost to the development project—not counting prototype testing and delays due to redesign. Faced with long delays, firms are questioning the role of prototypes and are searching for cheaper, faster alternatives. Wander the halls at Honda, at Sony, at Motorola and you will hear engineers discussing new concepts like "one-pass design" and "virtual manufacturing."

The concept of "virtual manufacturing" is employed by Alex Pentland and John Williams of MIT to describe the process of using computer simulation to interactively explore the "space of valid designs."[35] This approach to design "integrates the interactive character of pencil and paper with the ability to quickly perform dynamic and structural analysis." This integration not only leads to better designs by providing the capability to generate a large number of alternatives, but it also significantly shortens the time between creation of a design concept and analysis to validate the idea—allowing the designer a means to "better explore the space of physically valid design."[36]

In some industries, such as electronics, computer simulation and modeling skills have progressed sufficiently that, through virtual manufacturing, prototypes are rendered obsolete. In motor vehicle development, CAD systems have obviated the need for clay models and multiple prototype models. General Motors, for instance, "only build(s) a clay model to validate their aesthetics," according to Vice-Chairman Donald Atwood.[37] High-

[35]Alex Pentland and John Williams, "Virtual Manufacturing," NSF Engineering Design Research Conference Preprints, pp. 301–16, Amherst, MA, June 11–14, 1989.

[36]Ibid.

[37]John Bussey and Douglas R. Sease, "Manufacturers Strive to Slice Time Needed to Develop Products," *The Wall Street Journal*, February 23, 1988.

resolution, 3-D graphics monitors can now be used instead of models to validate aesthetics. Other firms are employing simpler prototypes to make changes cheaper and faster. In the development of a CAT scanner, Siemens built simple shells of the product, omitting the interior electronics, for analysis by marketing to get early feedback on the design at a fraction of the cost.[38] In computer design, firms often use simpler circuit boards for prototypes to avoid the delay and expense of integrated circuit fabrication; Commodore took this approach in the design of the Amiga 1000 PC.[39]

To become time-based competitors, firms must reassess the value of prototypes in the development cycle. Prototypes should be built only when required by regulation, as with an emission or crash test, or when a computer model would be inadequate. Prototypes are warranted basically to test for things that cannot be modeled. In many cases, however, prototypes are built because "that's the way we've always done it." With sequential, "over-the-wall" product development, prototypes are used as a communication tool—an object that manufacturing and marketing can inspect. Since the prototype is often the first look at the design for marketing or manufacturing, major specification changes that lead to time-consuming redesign occur at this late stage. Prototypes also serve as a design crutch for designers who use a trial-and-error ("build it to get the bugs out") approach to design. These practices are evidence that the designers do not expect to get it right the first time. Old habits die hard, and they are expensive while they live.

Teamwork

The JIT approach to new-product development places a much higher premium on teamwork than the conventional sequential approach, primarily because of the radical differences in information flow. In the sequential, "over-the-transom" approach, teamwork is relatively unimportant; information tends to flow one-way, like water in a pipe, from design to engineering and from engineering to manufacturing. Communication is mini-

---

[38]Nhora Cortes-Comerer, *IEEE Spectrum*, May 1987.
[39]Paul Wallich, *IEEE Spectrum*, May 1987.

mized, "functional parochialism" is allowed to operate unchecked, and time-consuming rework cycles are common because design errors are detected late in the process, far downstream. As a firm moves toward overlapping activities, however, information flows become more complex, and teamwork becomes increasingly important. In an effective, cross-functional team, information processing, instead of being a one-way flow, is more like distillation: design information is churned and refined through feedback from the other functions. The objective is the early detection of errors that is made possible through early release of information and early feedback on design quality from the other functions—engineering, manufacturing, marketing, and so forth. To achieve this, all the functions that play a role in product development must be represented on the project team.

The literature on team construction, motivation, and leadership is extensive and growing. In Chapter 6, John Bailey specifically explains the construction of new-product development teams at Honeywell and explains how different functions assume leadership roles in different phases of the project. Additional insight is provided by Dean and Susman's work on organizing for manufacturable design,[40] and Hackman's research on the design of work teams.[41] The project team leader's role is key. He is an integrator who must understand all the roles that different functions play in product development, and, at times, must referee the play. At other times, the leader must be the chief communicator and buffer between the project team and top management.

Automation

Automation should be viewed as the culminating, not the initiating, activity in efforts to reduce setup times. Similar restraint should be applied in a firm's attempt to crash time in product development. Automation has a seductive, quick-fix ap-

[40]James W. Dean, Jr. and Gerald I. Susman, "Organizing for Manufacturable Design," *Harvard Business Review*, January-February 1989, pp. 28–36.

[41]J. Richard Hackman, "The Design of Work Teams," in *Handbook of Organizational Behavior*, Jay W. Lorsch (ed.) (Englewood Cliffs, N.J.: Prentice-Hall, 1987).

peal to impatient managers with short-term goals. Managers in the United States, who contemplate rapid automation of the development process, however, should heed the lessons learned in manufacturing.

When JIT first emerged from the Orient, many U.S. manufacturers responded by rushing headlong into automation. These executives argued, "JIT won't work here; our workers and our culture are different. We can use our technological edge and leapfrog the Japanese in manufacturing." This rationale provided the impetus for an "automation crusade" in which firms threw dollars on the factory floor to purchase robots, automatic transfer machines, automatic guided vehicles, and totally automated warehouses. To their dismay, most automation-bent firms fell further behind the Japanese. In their rush to automate, these firms left unchanged the underlying, outmoded production processes; they spent huge sums of money to hardwire an inefficient process. Manufacturing was changed, but not unscathed.

Likewise, in their rush to come up to speed in new-product development, U.S. managers should approach automation warily. A firm's information process and project teams should be functioning smoothly before it embraces automation. Automating a sequential, batch process of design can, admittedly, quicken certain tasks, but it is unlikely to bring development time up to the level of global competition.

This is not to suggest that automation is not critical to accelerating new-product development. To the contrary, CAD and integrated CAD/CAM systems are as essential in today's design shops as word processors in the office. Work is proceeding on integrated knowledge-based CAD systems that improve design productivity and quality for plastic injection-molded parts and increase the likelihood of having first-pass manufacturable designs.[42] As mentioned earlier, computer simulation and high-resolution graphics can shrink design time and improve design

---

[42]H. Hanada and L.J. Leifer, "Intelligent Design System for Injection Molded Parts Based on the Process Function Analysis Method," NSF Engineering Design Research Conference Preprints, pp. 597–612, Amherst, MA, June 11–14, 1989.

quality, and they have the potential to eliminate costly, time-consuming prototypes. In electronics, software used in testing circuit board designs is taking large chunks of time out of the design phase and the testing phase. Automation alone, however, will not generate the desired development speed if problems lie within the process; managerial skill, rather than money, is the factor needed in the new-product development game.

Principle 3: Solve Other Problems that Arise

In product development, as with JIT, smaller batch sizes spark the outbreak of other problems, like small brush fires, that management must solve if cycle times are to be reduced. The problems are analogous to those that arise in manufacturing—design quality, layout and information flow, worker involvement, and supplier relations—and, as in JIT, these are problems that predated the process change but were obscured by large-batch information transmission. Several of these problems were addressed in the activities to reduce changeover times.

Information Flows and Layouts
In conventional batch manufacturing, when components are produced in large batches, functional layouts are rational because they take advantage of processing efficiencies. Similarly, in the typical Western new-product development process, the product moves through functional processing centers—work on designs is carried out sequentially and then shipped on to the next department—designers can be in one area, product engineers in another building, and so forth. With one-way, "over-the-wall" communication, the functional, job-shop layout can work, but it does little to promote speed. As the firm moves to overlapping activities with information transferred in smaller batches, work groups must move closer together, in space and in time, as is demanded by JIT. To function as a cooperative team, the members of the project should be moved out of their conventional departments and into a layout that will support continuous communication. For example, in the development of the Proprinter, IBM pulled about 100 people together for the project from all functional areas. According to Charles Rogers, the proj-

ect manager, the group "lived in the same complex, away from everybody else, dedicated to one project."[43] Brunswick Corporation took the unusual step of moving its outboard motor factory from Stillwater, Oklahoma, to an engineering facility in Wisconsin to speed product development by stimulating face-to-face communication between the functions.

Quality

This is a key issue because, in the United States, a lot of the time wasted in new-product development is due to quality problems—getting it wrong the first time! Designs, produced in batch, turn out to be deficient typically because the market requirements were inadequately understood or manufacturing capabilities were misjudged. In the design phase, as many studies have found, there is an overemphasis on product performance and not enough emphasis on manufacturability and cost.[44] Isolation and compartmentalization of the design function increases the likelihood that the designer is given the wrong objective or misinformation about product specifications and that the subsequent design will be rejected.

If new-product development is restructured to follow the JIT model, quality problems may actually increase in frequency, but not in severity. As the lot size of information shrinks and the transfer rate between designers and other team members accelerates, small flaws are detected early in the process. In the long run this improves the process because the team can quickly determine that the emerging product design is veering slightly off specifications. Corrections are made before the problem magnifies.

With intensified communication, designers are integrated into the project group. Consequently, as team members from marketing see the design coalesce and take shape, they can provide quick feedback to keep the design congruent with customer requirements. Process engineers monitor the development of

---

[43]Nhora Cortes-Comerer, *IEEE Spectrum*, May 1987.

[44]D. Whitney, T.L. De Fazio, and others, "Tools for Strategic Product Design," NSF Engineering Design Research Conference Preprints, Amherst, MA, June 11–14, 1989, pp. 581–95.

components for early detection of mismatching or other assembly problems.

The JIT model is likely to increase the time required for the first-pass design because it will uncover many small glitches before they become large problems. The goal, however, is to eliminate the waste of rework and get the product right the first time. Total design time, therefore, is dramatically reduced; JIT produces a higher-quality product in less elapsed time.

Supply Relationships

As batch sizes and inventories are reduced in JIT, material flow must be synchronized to ensure a continuous supply of defect-free components between work stations. To further compress time, shipments are made in small quantities from the supplier to the manufacturer. Therefore, cooperative relationships with suppliers become important in reducing lead time.

Suppliers also play a critical supporting role in diminishing product development time. As Clark has found, the Japanese leverage their long-term relationships with suppliers, built through JIT, to obtain technical support in product design. Few U.S. firms are in a position to exploit this "two heads are better than one" edge because of communication barriers formed through years of arms-length dealings with their suppliers. However, JIT is changing that; walls are coming down and long-term commitments are being forged with suppliers.

In the United States, the Japanese "transplants" are leading the way by their efforts to build supply chains that are similar to those cultivated in Japan by the parent company. In auto manufacturing for example, transplant firms seek to use vendors as partners in the development of new components.[45] According to Jerry Benefield, president and CEO of Nissan Motor Manufacturing in Smyrna, Tennessee, "We aren't making decisions on the basis of the next 10-day sales report or the next quarterly earnings. Our decisions have to do with where we're going to be 5, 10, and 15 years from now. That is why Nissan

---

[45]James Welch, Peter Jueptner, and Joseph D. Blackburn, Working Paper, 1989.

works to establish long-term relationships with suppliers, *involving the suppliers in the design of parts.*"[46]

Under time pressure to develop new products, U.S. firms are building similar partnerships with suppliers. In the heavy-duty Class A truck market, product development is becoming a key dimension of competition. The market demands more-fuel-efficient trucks, and so the trend is to aerodynamic designs. As a key supplier of components to the truck manufacturers, Rockwell has had to develop new components more quickly. Historically, truck designs changed no more than once a decade; competitive pressure has cut that time in half. Within that cycle, the usual time to design a new component has been about two years. Dissatisfied with that, one of Rockwell's major customers, Freightliner, informed them that they needed a new plastic hood and fender assembly with less wind drag, developed in half the time. With teamwork between buyer and supplier, they were able to bring the new component on stream in 12 months. Closer coordination between functions and overlapping activities made it happen. For example, while Freightliner was approving a design, Rockwell would be incorporating changes and ordering the tooling the same day. They used a team approach to program management and simultaneous engineering to accelerate the decision process.

Employee Involvement and Teamwork

Teamwork is essential to effective new-product design. Barry Bebb states that "The Japanese recognize that manufacturing quality, cost, and schedule are largely determined by how well engineers execute design engineering. Manufacturing cannot produce competitive products unless engineering joins with manufacturing to design competitive products."[47]

AT&T and others are finding that the fastest companies form multifunctional teams—AT&T is using it with success in

[46]Jerry Benefield, "Address to Automotive News World Congress," Detroit, January 9, 1990.

[47]H. Barry Bebb, NSF Engineering Design Research Conference Keynote Address, June 1989.

product development. AT&T applied the team concept to the development of the new cordless 4200 phone. Typically, the development time for a new phone product at AT&T was about two years. The projects were also carried out like a relay race—or the "over the wall" concept. They formed teams and gave them greater decision-making and spending authority. For various phases of the project, the teams set rigid time deadlines and gave the teams the authority to do what was needed to meet the dealines. The results: Development time was one year instead of the usual two, with lower costs and increased quality.[48]

THE TIDE IS TURNING

There are encouraging signs that some U.S. firms are catching up in the global new-product development race. Significantly, and not coincidentally, the fast-closing firms are the ones that have already successfully implemented JIT manufacturing. Speed, in all activities, is the cornerstone of their corporate strategy. They, like the Japanese, have transferred the skills, honed in manufacturing, to new-product development. Exhibit 6 shows how leading U.S. firms recently have dramatically reduced product development cycles.

Manufacturers in the United States must continue to cut development time *and* cost. Each, alone, is not enough. Despite recent advances by Big Three automakers, Donald Smith of the University of Michigan predicts that, just to keep up, Detroit-based firms will have to reduce development cycles by a third and costs by half by the early 1990s.[49] Many observers feel that Smith is an optimist.

Just-in-Time provides an ideal template for managing a new production process for two important reasons. First, the focus of JIT is on time and quality. Second, management does not have to learn a new set of tricks; the techniques can be

---

[48]Brian Dumaine, *Fortune*, February 13, 1989.

[49]John Bussey and Douglas R. Sease, "Manufacturers Strive to Slice Time Needed to Develop Products," *The Wall Street Journal*, February 23, 1988.

EXHIBIT 6
U.S. Firms' Progress in New-Product Development

Firm	Product	Results
Honeywell Bldg Products	Thermostat	4 years down to 1 year
Navistar	Trucks	5 years down to 2.5 years
IBM	Printers	4 years down to 2 years
Hewlett-Packard	Printers	4.5 years down to 22 months
Northern Telecom	Digital switches	Reduced development time 20–50%
Motorola	Pagers	3 years down to 18 months
Brunswick	Outboard motors	Reduced development time 25–30%
Xerox	Copiers	4–5 years down to 2 years
3M	Microfilm readers	3 years down to 2½ years

learned first in manufacturing and then transferred to new-product development. This provides additional impetus for undertaking a JIT campaign in manufacturing: There are ancillary benefits for the other functions of the organization.

For firms restrained by dilatory new-product development processes, the writing on the wall is clear: the race is being won by the swift. The only alternative to accelerating up to speed is to drop out of the race. Shrinking new-product development time requires the same contextual shift in thinking on the part of management as shrinking lead time in manufacturing: A change in focus is required from cost performance to time. The greatest cost, by far, is the cost of being late to market.

CHAPTER 6

HONEYWELL'S TEAM APPROACH TO NEW-PRODUCT DEVELOPMENT

John Bailey

Editor's Note: John Bailey is vice president and general manager of Honeywell's Building Controls Division in Golden Valley, Minnesota. In this chapter he describes how Honeywell has used the team concept to speed new-product development.

The Building Controls Division (BCD) of Honeywell designs, manufactures, and markets a variety of energy management and environmental control products for commercial heating, ventilating, air-conditioning, and lighting markets. Honeywell's products regulate the environment of buildings—temperature, humidity, and indoor air quality—and conserve energy. The division maintains approximately 150 basic product lines but offers as many as 2,500 individual varieties of products. These vary from simple mechanical products, such as mechanical thermostats, to state-of-the-art microelectronic direct digital controllers. The division employs about 1,250 workers in factories in Minnesota and New Mexico, headquarters in Minnesota, and sales offices in 37 cities. BCD sells to original equipment manufacturers like Carrier, Trane, Lennox, York, and Cleaver Brooks and also sells through wholesale distributors to independent contractors. In all of the division's business, it sells products to companies that resell them to a third party, that is, two-step distribution.

THE CRISIS AT HONEYWELL

New-product development at BCD is a critical ingredient in its overall process for total quality (TQ) improvement. Honeywell's quality improvement process is based on the fundamentals of conformance to customer requirements. It is a system of prevention rather than detection that relies on two principles: prevent a problem rather than find it, and do things right the first time. Just-In-Time and statistical process control, thus, fall under the umbrella of a total quality improvement process.

The TQ process began in 1982, at which time Honeywell faced a crisis. Profit margins were eroding, smaller competitors were gaining share, and many customers were dissatisfied. Consequently, the decision to launch a quality improvement process was not difficult. Honeywell was being beaten by some of its smaller, faster competitors and was fearful of potential Japanese competition. The firm was compelled to do something different to conform to its customers' demand for quicker response. To address this need, BCD focused in on new-product development, which had been studied during the crisis period.

When the crisis hit, Honeywell utilized the traditional approach to product development. It was a sequential approach, with each functional area responsible for a portion of the development process. To start, the marketing and sales people talked to customers and learned what was needed from a variety of requirements that were identified. After talking to a broad sample of customers, these personnel filtered customer requirements and translated them into product descriptions. The tendency of the sales and marketing employees was to devise product definitions that would be all things to all people. They wanted flexibility and a full range of features, and they also wanted the lowest-priced product on the market. After this market research and product definition, marketing management requested that engineering design and develop the product.

When engineers received a new-product request, they first studied it to understand the requirements. They learned the meaning of the requirements and tested viable approaches. The engineers also sought opportunities to apply the latest technologies. Frequently, either the cost of innovation was high, or the producibility of the product suffered. Trade-offs during the design cycle were negotiated between engineering and marketing. When the design was complete and thoroughly tested, specifications were recorded, and the product was turned over to production.

Production first made its own study of the requirements to determine if the design could be built with existing tools and processes. If not, it designed new lines and ordered new machines. Production personnel wanted to be sure that they could deliver; experience taught them that cost problems would get them a bad review, but delivery problems would get them fired. Delivery was their number one priority. When production conducted a feasibility study, the first question always was whether the product could be delivered to the projected schedule. Thus, products often were modified to ensure that they were deliverable at the desired line rate.

This traditional sequential process works, but it has many built-in problems. Honeywell experienced some of those problems. If a customer provided a new idea, or some seemingly different opportunity existed in the marketplace, marketing felt free, and even compelled, to change the product specification any time during the development process. Because of this historical pattern, engineering and production managers tended to make allowances for slack time in the schedule changes. They actually planned for changes in the product, both in the process and in the schedule. Each time a change order occurred, management initiated a new approval process for the new specification. The schedule frequently slipped, and both the product cost and the program cost increased. In this process, moreover, the products proposed often could not be produced with the available resources. Product designs were sent back to R&D for redesign or for reprocessing of the parts. Marketing, in addition, was often asked to change the customer's requirement.

Honeywell's sequential approach to product development consumed two, three, even four years between the product idea and its delivery to the customer. New-product development, in fact, averaged 38 months from start to completion. This lead time was too long for the existing markets; it proved to be a major reason that Honeywell lost market share to its competitors.

Consequently, in mid-1984, Honeywell dramatically altered its approach to new-product development: it instituted a team approach. During the last five years, in which the new team approach has been used, the average time between product idea and customer delivery has decreased from 38 months to 14 months. Honeywell has completed 20 major products and many minor developments using the team approach. The company has institutionalized the Peters and Waterman "skunkworks" approach and given it credibility within the organization. Honeywell believes it has found a preferable mode of operation for new-product development.

Because of the new team approach to product development, Honeywell believes that it is a little faster and a little better than conventional competitors like Allen-Bradley, Johnson Controls, and some divisions of Emerson, which are in various stages of working with the team approach. Honeywell has not had to face an influx of foreign competition and believes that the threat of it has been staved off. Honeywell is still being beaten, however, by the small startup companies that have entered the energy-management field. These venture companies find a small niche and use solid-state technology to bring forth a quick product as a quick response to a specific customer need. Honeywell is not quite as fast as these companies because they usually are a specialty shop with a couple of entrepreneurs, a few workers, and some capital. Honeywell, however, believes it has more staying power.

THE BUILDING CONTROLS DIVISION'S TEAM APPROACH TO PRODUCT DEVELOPMENT

The team approach employed by Honeywell assigns members from each functional area to the development project. The team is responsible for the business plan, investment analysis, and project schedule. The team remains intact, "from concept to carton," that is, during the whole period required to develop a product idea into a deliverable article. As shown in Exhibit 1, the degree of individual member involvement on the team, however, varies by function based on the phase that product development is in.

EXHIBIT 1

Team Involvement by Function in New Product Development

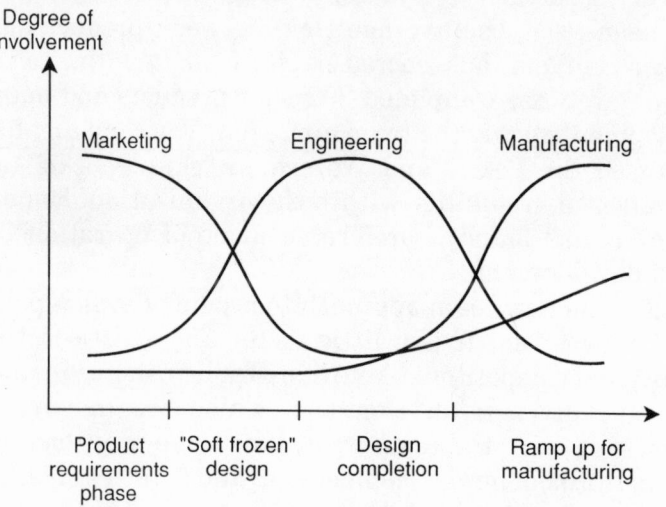

Marketing assumes predominant responsibility in the initial phase, the requirements phase. Team members from production and engineering, however, also visit customers and understand their needs right along with the marketing and sales personnel. The diversity that the former bring to the customer visit and their knowledge about engineering capability and production help define problems accurately and enable the team to devise a solid, enduring product specification.

While developing the specification, the team decides the key attributes on which the product is going to compete in the market. No one debates whether it is going to be a quality product: World-class quality is assumed and conformance to customer requirements is mandatory. Rather, the team decides whether the product will compete primarily based on flexibility (use in a variety of applications), on performance (functionality and features), on innovation (using new technology to do a job differently), or on cost. An initial decision on the competitive strategy keeps team members focused on priorities throughout the development. In this way, the new-product team does not encounter different functional areas vying to use their expertise in shaping the competitive strategy. It is decided up front. Downstream trade-offs are avoided by upstream agreement. When all requirements are fully defined and agreed upon through team consensus, and before the design begins, the specification is "soft frozen."

Once the specification is "soft frozen," the lead transfers over to engineering to complete the "design phase." The team revisits the specification only if something changes so much that designers cannot proceed. If a new idea surfaces, it goes into the next "generation" or the next product development. The product specification hardens as the design phase nears completion and as limits are placed on operating parameters, dimensions, and performance targets. Production involvement increases to assist designers in making sure the design fits existing manufacturing capability. If new capability is necessary, the team begins working on it at that point. Marketing's role decreases during this period, but it stays in the picture to participate in the trade-offs that occur during the design phase.

When the development testing is complete, the design is "frozen" and production takes the lead. This occurs without any surprises because team members from production have been involved since they called on customers to generate new-product ideas. Design engineering involvement diminishes, but its members remain on the team, committed to the project, in case questions arise or problems occur. Marketing's role increases at this time because its team members begin preparing for product introduction by coordinating test samples with customers, writing technical/training materials, and preparing promotional campaigns.

Honeywell's team approach to new-product development is analogous to a basketball game in that the players stay on the court throughout the game; everybody gets a chance to handle the ball, carry the lead, score, and play defense. Honeywell's former process was more like a relay race. Individual contributors carried the baton for a leg of the race. Problems occurred during the passing of the baton, resulting in many dropped batons and rough handoffs.

In contrast to its former sequential approach, Honeywell's team approach realizes a faster response to market needs and a clearer project objective among the team. It also develops better specifications because the team spends more time with a more diverse group on the front end. The team approach also has increased the amount of worker involvement, motivation, and commitment. Involving more personnel in the decision-making process has raised their level of contribution and productivity. The end result: Honeywell now makes better products and makes them faster.

FACTORS IN THE BUILDING CONTROLS DIVISION'S SUCCESS WITH PRODUCT TEAMS

The team-based approach to product development and production does not necessarily require substantial investment in capital or automation. It is not dependent on that. Teams, however, do benefit from having some computerized engineering, design, manufacturing, and modeling techniques, and other aids to the development process. Modeling techniques have proven useful in market research as well as plant layout. But such aids are not essential or mandatory for making the team approach work.

The greatest challenge to implementing team-based tactics is to adopt organizational and managerial styles that enable teams to operate and flourish. For instance, within the traditional marketing-to-engineering-to-production sequence, companies make numerous repetitions of requirements, definitions of requirements, product specifications, and product plans. Team tactics, by contrast, favor taking the time to arrive at a consensus on the definition of a new product before developing it. The essential difference is that the traditional approach builds in time to do everything over once or twice to achieve the right specification, while the team approach takes enough time to do things right on the first try. The greatest down side risk is that organizational "turf battles" may inhibit team effectiveness. See the following chart.

BCD's Team Effectiveness Factors

- Organizational parity
- Two-way communication
- 5–6 member teams
- Managing priorities
- Dispute resolution
- Simplified procedures
- Team support
- Training
- Supplier involvement

Parity on Teams

Honeywell discovered several basic prerequisites to successful implementation of the team approach. It found, for instance, that "organizational parity" is absolutely essential. Operating functions on the team have to be seen as individually important and equal. When Honeywell engineers teamed with marketing people, sales people, and manufacturing people, the team members did not have parity in compensation. Engineers used a time card reporting system and, therefore, had a compensatory status different from other members of the team. Consequently, Honeywell eliminated time reporting for engineers, and compensated everyone on the team in the same manner.

Marketing team members had a hard time during the initial process in sharing decision making with personnel from other functional areas. In the past, marketing had made all the decisions about customer requirements. Manufacturing people had a hard time recognizing that making customer calls was beneficial in the long run for making specifications. The BCD had to address these problems and educate team members accordingly.

Thorough and accurate marketing research at the outset is essential. There is a tendency for everybody to want to move forward too soon. The team members understand what is needed and want to proceed quickly. Taking additional time usually pays off. By spending additional time on product definition, the team becomes certain that it has defined the right product to fill customer needs.

Two-Way Communication

Among the new management characteristics that Honeywell developed to enhance the team process was uninhibited, two-way communication. Team members must be free to communicate both good news and bad news; management must feel free to communicate openly about the overall business of the organization, the total strategy, and the company's competitive situation. The channels of communication must be open so that everyone in the organization sees the whole picture and not just a piece of it.

Team Composition

The optimum number of people on a team varies with the complexity of the project, but usually is five or six. A team with fewer members lacks adequate diversity and representation from the functional areas. Finance personnel, while not full-time team members, play a role on every team. Also, the closer that all team members come to serving full time on the project, the better the team functions. Occasionally, people serve part time on two or three teams, but teams seem to work best when five or six people work full time on the projects.

Teams of more than six members have a harder time reaching consensus. Some Honeywell projects, for instance, have tried to use a 12-member team, but found more success by dividing programs into two pieces and then merging them at the appropriate juncture in the development process. Development of complex, sophisticated products, however, can warrant the use of large teams. A large team, for instance, developed a direct digital controller for large industrial process burners and boilers. This controller contained some 600 parts and as much software as an Apple computer. For projects of this kind, Honeywell forms a consensus team that has more than one lead person, such as a lead engineer, the product manager, and the chief process engineer, to organize the project team and make the consensus decisions.

A senior person always leads a development team and is

selected depending on what product attributes are highlighted. For example, if the product is to compete based on manufacturing flexibility, a senior person in manufacturing leads the team. If the product is to compete based on technical innovation, a senior person in engineering leads. Ideally, the team leader stays with the project all the way through product development. Sometimes, however, changes in the competitive strategy dictate a change of the team leader.

Honeywell found that technical experts, masters of their science and great individual contributors, usually disliked being on teams. Their skills were essential to development, however, so management classified them as "staff consultants" and allowed them to work concurrently with several different teams. If a team encountered an area that necessitated in-depth technical expertise, it enlisted help from one of the staff consultants.

Setting and Managing Priorities

New-product projects originate with the strategic planning teams in the BCD's business units. Each business unit has a strategic plan that lists its project preferences. A brief description of the intent from the strategic planning team identifies the segment of the market to target. Functional heads of marketing, engineering, and production assign their team members based on the intent of each project.

Program funding—authorizing capital and production equipment—is approved up front at the same time that specifications are approved. If milestones are maintained, the project is funded until capital appropriation time, and it is understood from the outset what capital requirements will be. When the time comes to order tooling, there is another capital appropriation cycle. After the initial approval, this is the only other approval cycle for the project.

The list of priority projects in strategic plans are reviewed three or four times a year. Priorities are reordered based on factors such as return on investment, urgency of customer need, and degree of competitiveness to be achieved. Projects that promise cost reduction gain support from reviewers; however, that factor is not as important as the prospect of losing business or missing a market opportunity if the product is released too late. Setting priorities demands a customer focus that is based on a feel for the business and a knowledge of the customer.

When a project is not a priority, the team leader understands the whole picture and why the product is not priority. When the team makes its milestone plan, it knows where the project stands on the priority list. Team performance is measured against its milestones regardless of project priority. Teams are not penalized for lacking priority or when priorities change. If priorities are reordered by a trade-off decision, the teams affected revise their milestone plans.

Dispute Resolution

Management uses monthly project reviews. Team leaders attend the review sessions and present their projects. If there is substantial disagreement over some aspect that cannot be resolved by the team, that problem is raised with all the functional heads present. The review team either resolves the issue on the spot or the functional heads later meet with the entire team involved and resolve the matter. In either case, disagreements never last more than a month.

Simplified Procedures

In its 103 years of operation, Honeywell had accumulated many procedures, policies, and traditions. Most of these did not surface until someone tried to do something differently. Management discovered that the engineering manual, which contained 100 years' worth of procedural rules, was about 4 inches thick and included 46 different sections. Rather than aiding the process, the manual retarded it with superfluous checks and balances. Honeywell did not discard this manual, but simplified, updated, and modified it. A concise 20-page, loose-leaf notebook also was developed as a current procedural guide. No one read the antiquated manual; now a usable notebook leads team members through the product development process.

Support of Teams

Honeywell also discovered that recognition and support of its teams are quite important. Each team must feel it possesses decision-making power. If the company envisions a sizeable project that has the potential to be a tradition-breaker, upper management assigns a senior executive to be the champion of the team. This person attends occasional team meetings to assure the team members that they have company support. The company always recognizes major milestones in a project by sponsoring dinners, awarding plaques, and expressing recognition in other ways. This practice motivates team members toward continued diligence in pursuit of their project goal. In addition, the BCD encourages calculated risk taking by the new-product development teams. To do this, the company removed reprisals and informed personnel that it was acceptable, and often necessary, to take a risk and fail.

Training for Teamwork

Honeywell also found that initial training for the team members was essential. Team members must possess the skills needed to make the decisions that they are asked to make. Some team members may have never participated in a decision-making process; they required training in problem solving and decision making.

Involving Suppliers

To achieve its current development time of 14 months, Honeywell used incentives with its suppliers to entice them to reduce their lead times for delivering components needed for new products in production. Honeywell also used incentives, instead of pressure, with suppliers to reduce the cost of these components. Moreover, the firm dramatically reduced the number of suppliers and began involving some vendors in development projects. When it started the quality improvement process, Honeywell used 6,000 vendors; today, it uses 650. Ten percent of those

650 are certified vendors whose goods are accepted without testing. These vendors work closely with Honeywell's development teams. Representatives from vendors become involved early in the design phase and review the program with the team at each stage.

BCD'S CONTINUING CHALLENGES

By using a team approach in new-product projects, Honeywell reduced average development time from 38 months in 1983 to 14 months in 1988. The total people time (the amount of people months per project), however, has gone down by only 5 to 10 percent. Despite the substantial reduction in project duration, increased team time is required both up front and throughout the process. Honeywell optimizes people time by simply not doing things over and by making few changes once the project is underway. The real payoff is not in decreased cost, but in increased responsiveness to market opportunities, a better finished product, and more satisfied customers.

New-product development time at Honeywell has not been pushed to its limit; the wall has not yet been hit. Although management believes that 14 months is a worthy achievement, it also recognizes that the competition is not getting any easier. Managers are exploring additional ways to reduce the average development time to under $13\frac{1}{2}$ months and then down to 12 months. Management does not know what the ultimate limit may be, but remains convinced that it has not been reached. Honeywell strives, therefore, to keep its momentum going toward time reduction, because it realizes that the market becomes more competitive and more global every day.

CHAPTER 7

XEROX CORPORATION: A CASE STUDY IN REVITALIZING PRODUCT DEVELOPMENT

Mohan Kharbanda

Editor's Note: In this chapter, Mohan Kharbanda, who holds the position of Manager, Business Strategy, at Xerox Corporation, provides a view from the inside of how Xerox recognized that slow new-product development was weakening its competitive position and the steps it took to come back up to speed. Xerox is a $16 billion company that has earned international renown in multiple industries. In recognition of its significant and innovative approaches to quality, Xerox Corporation was awarded the prestigious Malcolm Baldridge Award in 1989 by the U.S. Department of Commerce.

XEROX IN RETROSPECT: THE RISE AND IMPACT OF COMPETITORS

Xerox created the plain-paper copying industry in 1959: the year Xerox introduced the 914 copier. Prior to 1959, the most popular office copying techniques were relatively messy, inefficient wet processes; carbon paper was widely used. The 914 changed all that and transformed the Haloid Company of Rochester, New York into the Xerox Corporation. The company's—and the world's—first plain paper copier, the 914, was one of the most successful new products in corporate history. For the next 15 years, Xerox was the victim of its own early triumphs.

Xerox had such a stranglehold on the copier market throughout the 1960s and early 1970s that little attention was paid when the International Business Machines Corporation and the Eastman Kodak Company began marketing high-speed copiers, the most lucrative part of the market. Nor was there much concern when the Japanese began to offer small, inexpensive copiers in the mid-1970s. In fact, Xerox ignored this product segment until recently.

Xerox of the mid-1970s was a bureaucratic company in which one function battled another, and operating people constantly bickered with corporate staff. Disputes over issues as relatively minor as the color scheme of machines had to be resolved by the CEO. The result was painfully slow product development, high manufacturing costs, copiers that were hard to service, and unhappy customers.

Then, in the mid-1970s, the Japanese camera makers entered the low end of the light-lens copier market. They used aggressive pricing to gain a foothold and proceeded to gain a sizable market share. This strategy was similar to the one the Japanese used so successfully in automobiles, cameras, home appliances, calculators, and watches: compete aggressively in the low end of the market and then gradually move to the more profitable mid and high ends.

The Japanese strategy worked well. Through the late 1970s, Xerox saw its market share erode at an alarming rate. However, the problem was masked for a while because of the rapid growth of the entire copier market. Years of immunity to international competitive vulnerabilities had created a kind of "arrogance of entitlement": an abiding belief that whatever was going to be invented or manufactured, America would set the standard. America would do it first and do it best.

In fact, when Xerox recognized the influx of Japanese competition into the copier market, the company sponsored a conference solely to determine if the Japanese posed a real threat. The conference scientists and speakers concluded that the probability of a company manufacturing a copier—in Japan or in the

United States—at a lower cost than Xerox was very slight. Any new company entering the market, the conferees concluded, would lose money because the cost of entry would be prohibitive. Only a few weeks later, however, Canon launched a low-end copier (the Model NP210) that Canon sold at a price lower than Xerox's manufacturing cost. The conclusions reached at Xerox's conference had proved to be dangerously incorrect.

By 1980 the problem had become readily apparent. The company had to rapidly evolve into a world-class organization if it were to effectively compete with worldwide competitors in the global market.

BENCHMARKING: THE GENESIS OF CORPORATE RENEWAL

Xerox began to respond by assessing its corporate strengths and weaknesses, as well as those of its competitors. One of the most important management actions occurred in 1980 when the company instituted a formal *benchmarking* process to identify the successful practices of top competitors in each of Xerox's operations. Forms of benchmarking have been used in industry for years. Early in this century, Walter Chrysler would tear apart one of each new Ford model as soon as it came off his competitor's assembly line at the beginning of the model year. Chrysler sought to determine what components went into the car, how much they cost, and how they were made. Armed with this information, Chrysler had a better understanding of the strengths and weaknesses of the major competition.

Competitive benchmarking at Xerox is a tool used to identify industry performance standards. It provides insights into how these performance standards can be achieved or exceeded and how to develop internal action plans. Most importantly, benchmarking is an ongoing learning experience for the firm as a whole. At Xerox, competitive benchmarking looks both inside and outside the reprographics industry.

What did Xerox learn? The firm got quite an education during the process and is a lot smarter as a result. First, Xerox learned that it took too long to develop new products. Moreover, the products cost too much and did not fully satisfy customers' requirements. In fact, Xerox discovered that the Japanese were selling their small machines for what it cost Xerox to make its machines. The benchmarking process enabled Xerox to learn how its competitors managed costs and how those cost-reduction tactics might be replicated. The firm learned that it needed to overhaul the way the business is managed: to compete successfully, Xerox had to be driven by its customers and its competition.

Internally, Xerox discovered that one of its problems was organization. Xerox had a matrix management structure in the product development group. This means that a product development project must flow through separate functions—product planning, design engineering, manufacturing engineering, and service engineering—each operating independently, almost in a vacuum. Significantly, no individual had clear responsibility for the end product.

The matrix structure had been installed to prevent errors. Instead, it had the unintended effect of blocking product delivery. With so many different functions involved, slowdowns were inevitable. Time-consuming committees were needed to address cross-disciplinary issues, and every product problem was cross-disciplinary. In addition, there was constant need to review and gain concurrence across the different functions in the matrix. The product development cycle was so long, in fact, that products sometimes became obsolete in midstream because the market needs had changed.

Xerox began the process of renewal by dismantling the matrix organization and the bureaucracy accompanying it. The product delivery organization was restructured into product delivery teams headed by one manager, called the chief engineer. The chief engineer is held totally accountable for a product development project, including quality, cost, performance, and schedule. The chief engineer manages all the design group, the model shop, and the pilot plant. He has complete authority to modify the development schedule and to make go/no go decisions along the way.

Not until 1984, however, did Xerox make its way back into the market. By then, Xerox had made significant progress in product quality and cost. Before 1980, for instance, Xerox copiers averaged 1.2 defects per unit after installation at the customer base. Today, on average, only 6 out of 100 Xerox copiers ever have a defect. When Xerox began its benchmarking process in 1980, it targeted several objectives to be met over a five-year period. By 1985, Xerox was to reduce its product costs and product development time by 50 percent, and to increase its product quality by over 90 percent. Within five years, Xerox was to regain its former status in the copier market and, equally critical, to be the industry leader in customer satisfaction.

As soon as Xerox matched competition on quality and cost, the rules began to change again. In fact, Xerox did reduce manufacturing costs by 50 percent—but so did the Japanese. Xerox improved product quality, but the Japanese already had matched that quality. Compounding the problem was Xerox's inability to reach its goal of reducing development time. The new battle was being waged over time to market.

Time to Market

The new-product development process at Xerox begins at the product planning stage. In this stage, the firm seeks to identify market trends through customer surveys and focus groups by having potential customers "build," conceptually, the ideal copier. Customers, through their responses, identify trade-offs they are willing to make, such as sacrificing some speed for additional paper-handling features. The result is a product design that matches appropriate technologies with customer requirements. For instance, Xerox learned that users of one copier model disliked having to shut off the machine to add paper. Consequently, the recently introduced 1090 copier comes with two paper trays that can switch automatically when one is empty.

Xerox's foremost challenge remains that of improving its product development process and decreasing the time it now takes to bring product releases to market. The prime reason is that technology in the copier industry continues to change rapidly. The technology of facsimiles, laser printers, and multifunctional products that will be on the market in a few years indicates that the copier of the future will be more than just a copier.

It will be an "intelligent" copier; that is, it will function as a copier, a facsimile machine, and a workstation printer. It will receive input from any location and, eventually, will be integrated into a desktop publishing unit. These capabilities will foster other advanced technologies. Xerox must shrink its product development time if it is to participate in the industry's growth.

Consider two products that appear in the marketplace in 1992, one from a company with a two-year development cycle and the other from a company with a three-year cycle. The two-year cycle product incorporates the technology of 1990; the other, the older technology of 1989. The result is technological leadership, even though nothing new has been invented. This kind of disparity can dictate success or failure in new, emerging markets.

MAGNITUDE OF THE NEW PRODUCT PROBLEM

Xerox has used traditional motivational tools to increase productivity by 5 to 10 percent annually. But the firm's benchmarking efforts reveal that these gains may not be sufficient in the 1990s. In product development, Xerox found that its Japanese competitors were using a fraction of the development time to launch new products in the marketplace. Matching this rate was essential to the company's goal of accelerating the rate of new product introductions and of reducing product development costs. Specifically, Xerox found that unless it reduced its new product development and introduction cycle from 36–48 months to 12–24 months, the Japanese and other competitors would out-innovate and outperform the firm. This finding was based upon the following observations:

- The Japanese believe that in a world of high technology, rising yen, and improving Western emphasis on quality, they need to compete not on quality or cost, but on innovation and schedule speed. The evidence so far indicates that they are successful. The leading Japanese firms have shown that they have the capability to launch three products to Xerox's one.

- The leading Japanese new-product developers have an advantage in time and productivity over Xerox. Exhibit 1 displays this dual advantage. With a 50 percent time advantage and, at the same time, a 30 percent manpower advantage, the Japanese new-product development efforts were almost three times as productive. The difference in development time between Xerox and its Japanese competitors, therefore, is on the order of 1 to 2 or 1 to 3. This huge imbalance requires 200 to 300 percent more productivity, not the 5 to 10 percent that has been achieved.

- To shorten product life cycles further, the Japanese have revised their product delivery processes through the use of overlapping design phases, fewer prototype iterations, and holding design schedules firm by treating quality and cost as secondary variables to be traded off. Their goal is to reduce the design schedule to 12–18 months.

EXHIBIT 1
Japan's Productivity Advantage in New-Product Development

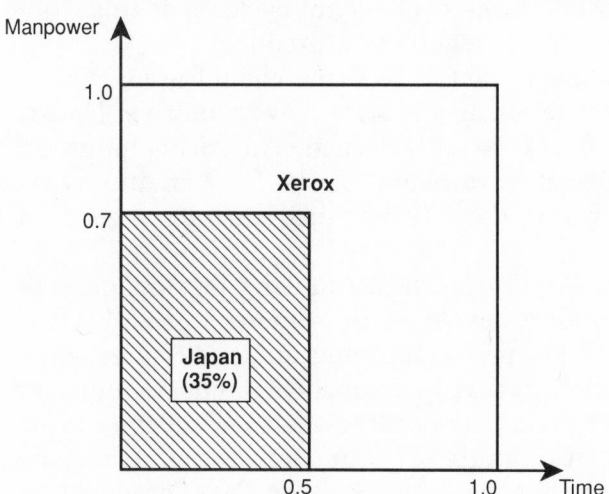

The response to the design schedule challenge requires nothing less than an internal revolution in all phases of product design, production, and distribution. Some of the steps Xerox is taking to achieve these changes are described below.

The Kaizen Approach to New-Product Development

During the mid-1980s, Xerox collaborated with its own Japanese copier venture, Fuji Xerox, to implement actions to close the product delivery gap. One of these practices is *kaizen*, that is, the incremental and continuous technology transfer process. Short delivery cycles and incremental transfer of technology into new products are strong practices that, if used in tandem, make any company a formidable competitor. In general, kaizen product planning seems to be the norm for Canon, Hitachi, Mitsubishi, Honda, and others.

Kaizen is a primary strategic element in new-product competition in a maturing industry. Within manufacturing, our problems occur not in the early stages of an industry but once it matures and settles into a cycle of gradual improvements, year after year. In the long run, being successful in this process, known as the maufacturing-development cycle, is as important as a firm's ability to create wholly new products.

This does not mean that U.S. firms should give up their ability to create new products and start new industries. That's a great strength that should be retained. American industries have always been good starters of projects. Now, in an increasingly competitive world, these firms must learn to be good finishers as well.

The significance of kaizen in accounting for Japanese advantages in new-product development is shown in Exhibit 2. In developing automobiles, personal computers, and copiers, Japanese companies begin not with clean-sheet products, but with variants of existing products or off-the-shelf technologies. Variants are members of a family of products to which a company makes incremental changes as it proceeds. New products are

EXHIBIT 2
Kaizen versus Conventional U.S. Approach to Product Development

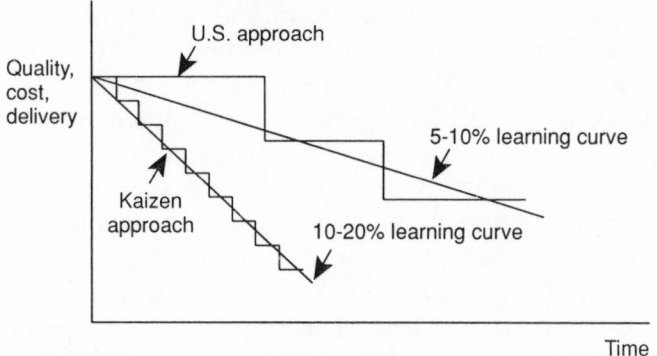

launched quickly by this method. The company then makes innovations in technology, features, or cost reduction in the successive generation(s) of those products.

MOVEMENT TOWARD PRODUCT VARIANTS
AND PARALLEL DESIGN

Competitive disadvantages caused Xerox to revamp its product
planning process. No longer is each product planned individually; instead, whole series of products are planned. If a product
does not result in variants, it is not a successful product. A product also must have commonality with many other products so
that advances in component technology will benefit multiple
products. Xerox now redesigns individual parts in a product instead of investing blocks of time to redesign the whole product.

Most U.S. businesses perform product development steps sequentially. They proceed from product preconcept to concept
planning to concept design. A great deal of time is invested in
developing the product concept, designing the prototype, testing
it, designing for manufacture, and launching. Authorization to
proceed to the next phase can take six months. Another six
months passes to secure approval and agreement on manufac-

turing costs, specifications design, and so on. At Ford, this process takes roughly six years.

In the past, Japanese companies worked sequentially, one step after the other. After the yen shock in 1985, however, the Japanese began performing product development activities in parallel. This practice, called the *Z curve phenomenon*, enabled Japan to reduce substantially the product development cycle. Honda, for example, undergoes one phase in contrast to several phases at GM and Ford. Before all specifications are determined, the design process begins, and changes are made as needed, even in the middle of the process. A prototype is assembled and tested before the design phase is completed. Before design testing is completed, the manufacturing line is set up and tooling begins. This Z curve process creates a substantial amount of bureaucratic and organizational turmoil and would work well only in U.S. firms that have the discipline to manage the process. The Japanese make it work and, as one result, significantly reduce their development time.

The Japanese also reduce development time by decreasing the number of prototype cycles. American businesses use the first design prototype for the engineering feasibility model. Another design is completed for use by other areas that have problems in their designs. Only then are manufacturing blocks, and, later, the launch blocks, formulated. This process demands up to five years. Currently, Japanese firms complete one engineering and one manufacturing prototype. Then the product is launched. They reduce the number of prototype cycles by designing accurately the first time. They use computer-aided engineering (CAE) to simulate subsystems and to integrate subsystems at the computer level. Computers in the design process are vital for optimizing time.

Xerox is reassessing its sequential design process and is reducing its number of design prototypes. The sequential processes now in use at Xerox decrease efficiency in product development. Honda succeeded in moving from motorcycles to automobiles partly because of its Z curve process for new-product development.

In its benchmarking, Xerox also discovered, at firms such as Honda, significant differences in the way U.S. and Japanese

EXHIBIT 3
Technology Transfer United States versus Japan

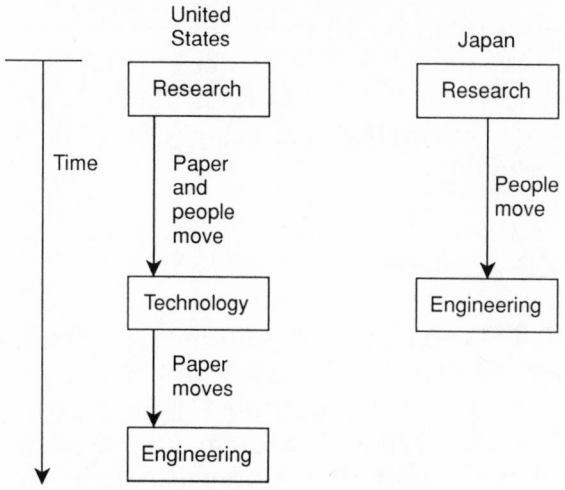

firms transfer technology within the organization. Honda, for example, moves product designers from the research labs into the business units as needed. In contrast, Xerox—like other U.S. corporations—passes the design concept, *not* designers, from the central research unit to an intermediary where the design concept is processed and ideas are negotiated. As shown in Exhibit 3, the critical difference is that, in the United States, the transfer medium is primarily paper; in Japan, people are used to propagate technology. In addition, Xerox, like many other U.S. firms, added an intervening organizational unit to evaluate research and translate the technology into an engineering application. This does not imply that, in Japan, the technology-related activities are not performed, just that they are performed in either the central research or engineering groups. In Japan, people are easier to transfer as "technology carriers" because the same grade/job structures exist between research and engineering and the reward mechanisms tend to be identical, so there are few disincentives to move from one part of the organization to another.

When a company moves the paper rather than the people, it

loses one to three years in matching new technology to the product. On the other hand, if a company uses an intermediary unit, it loses another one to three years. As a result, some technologies at Xerox never come to fruition, while others arrive too late to be useful. A competitive company cannot afford to lose these years in product development. With products such as PCs and copiers that have two-year life cycles, such delays can doom an internally developed technology.

NEED FOR QUICKER RESPONSE TIME

Businesses in the United States typically seek improvements in quality and procedural efficiency but neglect environmental changes needed to parallel those currently used in Japan with great success. Hitachi, a large $20-million company that withstood Japan's yen shock with relative ease, provides one example of such environmental factors. Hitachi's organizational chart lacks the names of some of the company's most influential personnel. The company practices team management through which these leaders simply stay in their positions while they oversee a specific project. One important reason that Hitachi is strong and successful is its team approach to management.

This organizational concept differs significantly from firms in the West, whose organizational concepts evolved from the Catholic church and the military, where specific responsibilities are clearly delineated. Job boundaries also are clearly drawn. In contrast, the Japanese delineate work responsibilities in ways that leave boundaries fuzzy. Japanese businesses define jobs by the personnel performing the work. The result is that employees work on the problem rather than just on what a job description dictates. The Japanese also bring in workers to oversee a project. A "hawk organization" is formed specifically to complete the project and, therefore, disbands quickly. Hawk organizations frequently accelerate the speed of response.

Japanese and U.S. businesses differ in other practices as well. Strategic planners in the United States typically remain in that function for the duration of their careers. They move from division to division, receive promotions, and eventually retire. In Japan, however, workers seldom leave their divisions. They move from job to job within the same division—from finance to production to procurement to personnel to sales and so on. Fifty percent of section managers may be rotated each fiscal year and the next-level managers may be moved every two years. As a result, managers acquire detailed knowledge of every function within their divisions and solve problems with great speed. Job rotation, overlapping responsibilities, and a very high emphasis on generalists are some of the reasons why Japanese firms adjust so quickly to shocks, such as energy crises, currency fluctuations, advances in production technology, and other major discontinuities. Western businesses, by contrast, form quality improvement teams and invest as much as six months analyzing a problem and seeking solutions. Still, the situation may not change because companies often fail to implement proposed solutions, and the response time continues to mount.

Japanese business practices increase the speed of response. When major upheavals occur in the marketplace—such as the yen shocks of 1973 and 1979, or a tariff barrier like that imposed against Canon by the European common market—Japanese businesses respond to them quickly. Xerox needs to improve its response time. For example, Xerox launched a project team in 1985 to overtake Canon because Canon's profits had leveled off due to the yen shock. It took Xerox a year to launch the project team and another year to develop solutions. By then, it was too late. Canon rebounded and is stronger today than it was before the yen shock.

Speed of response is a key factor in market success. Decision making, too, must be quicker in all components of the organization. Although Xerox is an improved company over what it was in 1980, it is not yet where it should be to remain competitive. Xerox continues discussions with other U.S. companies, including Ford and GM, who began a process of change five years earlier than Xerox because they faced Japanese competition that much earlier. Although these companies, like Xerox, now measure up to Japan in quality improvement, they still battle the problem of response time. Xerox has other fundamental changes to make as well, changes relating to its environment, the culture of the company itself, the tools of productivity, and the way in which Xerox thinks and plans.

As a result of all this, Xerox is now able to deliver products in half the former development time. Engineering productivity has more than doubled since 1985. For example, Xerox needed only two and a half years and 300 to 350 employees to develop its top-of-the-line 9900 copier, a high-speed machine. By previous standards, the project would have taken five years and in excess of 1,500 people.

After a decade of effort to develop the strategies and put in place the systems for Xerox to be a world-class competitor, it has significantly narrowed the advantage held by Canon, Ricoh, Sharp, and Minolta, stemming their advance in Xerox's market. This was recognized by the U.S. Department of Commerce when it awarded Xerox the Malcolm Baldridge National Quality Award in 1989. But for all of these efforts, there is more to be done, especially since the competition is still a target moving forward. Xerox may no longer be the company it once was, but it is not yet the company it wants to be.

CHAPTER 8

NEW-PRODUCT DEVELOPMENT AND MANUFACTURING COMPETITIVENESS: A HEWLETT-PACKARD PERSPECTIVE

Stephen Hamilton

Editor's Note: Hewlett-Packard Company is an international manufacturer of measurement and computation products and systems recognized for excellence in quality and support. The company's products and services are used in industry, business, engineering, science, medicine, and education in approximately 100 countries. HP, founded in 1939 and headquarted in Palo Alto, California, has 94,000 employees and reported revenue of $11.9 billion in its 1989 fiscal year.

In this chapter Stephen Hamilton, a strategic manufacturing consultant for Hewlett-Packard, describes how HP has used "stretch objectives" to integrate functions and time-compress new-product development cycles. He also describes how this concept has been applied to improve quality and manufacturing throughput times.

STRATEGIC OVERVIEW

Americans have slowly been awakening to the unpleasant news that, as a nation, we are losing our competitive edge. This decline has been well documented. Since 1970, American productivity performance has been dismal. Japanese productivity growth has been five times greater over the same period. In the automotive, steel, and other industries that had previously been considered America's stronghold, the numbers show a depressing similarity. America's share of world exports dropped precipitously from 26 to 18 percent between 1960 and 1988. America has rapidly moved from being a net exporting nation to a net importing nation. Although the United States still has the world's strongest economy, alarming statistics cast doubt that this preeminence will continue. As the President's Commission on Industrial Competitiveness reported,

> Competitiveness is the degree to which a nation can, under demanding and rapidly changing market conditions, produce goods and services that meet the test of international markets while simultaneously maintaining or expanding the real incomes of its citizens.[1]

Being competitive, or meeting the test of international markets, as the President's Commission reported, implies persuading the customer to choose a product over the competitor's. This should not be achieved through market protectionism or by purely superior marketing hype, but by providing greater value to the customer than the global competitors. Persuading the customer to buy the product is a simple strategy to state, but it cannot be achieved unless the reasons why the customers choose the product in the first place are well understood.

An effective business strategy is derived from the answer to that one question: Why will customers buy the products or services of one company rather than those of the competition? When analyzing any decision to purchase, whether it is a VCR, a suit, or a computer, four fundamental questions arise in the buyer's mind: what are the product's *costs, quality, availability,* and *features*? Across all product categories, customers buy products based on their perceptions of these distinguishing attributes. (See Exhibit 1.)

---

[1]*Report of the President's Commission on Industrial Competitiveness*, chaired by John A. Young, President and CEO, Hewlett-Packard Company, 1985.

EXHIBIT 1
Paradigm of Exclusivity

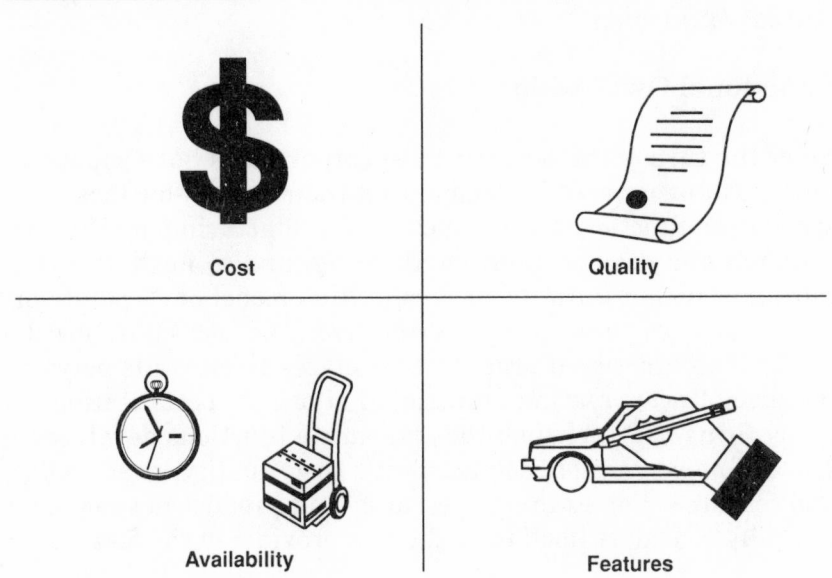

Cost	Quality
Availability	Features

Conventional wisdom has it that firms compete by distinguished performance on a single attribute, such as being the low-cost producer, or being the firm that offers products with the best features. This conventional wisdom was severely tested in the late 1980s by leading Japanese firms who exhibited the capability to offer products that excelled on all four attributes; these firms were quick to market with cost-competitive, high-quality products having all the features that customers desired.

To be successful in the 1990s, it is apparent that companies must be competitive on all four product attributes, not just one. Progress must be made simultaneously along all four dimensions. By pursuing the goal of availability, or timely delivery to the customer, the time-based competitor is secure in the knowledge that progress can be made simultaneously along the other dimensions. The firm does not have to give up one attribute for another. No trade-off is involved.

In pursuing this multiattribute strategy, however, a firm

must be prepared to deal, as HP did, with two formidable obstacles within the organization: (1) functional ownership and (2) mutual exclusivity.

Functional Ownership

Since the turn of the century, most corporations have supported functional ownership by segmenting their overall business into specialized functional areas such as manufacturing, marketing, research and development, engineering, and so forth. (See Exhibit 2.) Though initially successful, this model of corporate organization has some very severe drawbacks, especially for the 1990s. The functional organization places thick walls between the disciplines, creating "functional silos" of specialization. In many firms, HP included, these separate functions developed a sense of *ownership* of each attribute: cost, quality, availability, and features. For example, research and product development typically considers itself to be the sole provider of the feature set

EXHIBIT 2
Business Framework

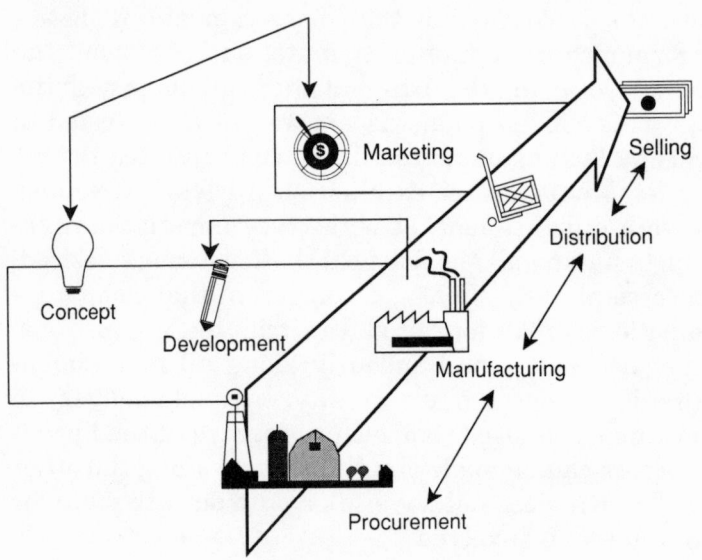

of the product, while marketing, on the other hand, assumes responsibility for product availability. Manufacturing traditionally focuses on cost containment and cost avoidance, and the quality assurance function focuses on quality. Independently of the corporation's goals and strategies the individual functions tend to focus on their particular attribute. As a case in point, a major food processing corporation stated that its strategy was to be responsive and flexible to its customers—that is, focus on availability. However, examination of the manufacturing capital spending plans over the past three years revealed that more than 80 percent of the capital budget was for cost-containment projects. In the majority of American corporations, the manufacturing function is managed solely on cost containment and cost avoidance independent of why the customer buys the product.

These functional silos are one of the most significant impediments to developing an interlinked view of the business. Hewlett-Packard realized that these functional silos must be broken down. The firm must move from a collection of suboptimized functions to an optimized whole in order to remain competitive in the 1990s.

Mutual Exclusivity

At HP and other corporations, the traditional view was that decision criteria—cost, quality, availability, and features—are *mutually exclusive*. In setting its strategic priorities, the firm must make trade-offs among these criteria. The process was viewed as a zero-sum game in which gains in one dimension are offset by losses in another. For instance, the "cheapest" item is rarely mistaken for the highest quality item. Or, to put it another way, the corner grocery store open 24 hours a day competes on availability; it is not often considered to be the cheapest source of products, or the one stocking the widest range, or even the highest quality items. In fact, pursuit of one attribute, such as cost, can cause a degradation in performance with respect to other attributes, if approached in the wrong way. For instance, many businesses approach the concept of product availability by keeping large finished-goods inventories, which clearly drives cost up, and in many businesses, notably the food industry, drives quality down. Because of this trade-off between attributes, this one dimensional outlook, therefore, can be called the *paradigm of exclusivity*.

Both of these entrenched beliefs—functional ownership of competitive criteria and mutual exclusivity—have, to a great extent, been broken at HP over the past decade. A paradigm shift has occurred based on the realization that trade-offs can be avoided and multidimensional objectives can be pursued, if managed properly. To begin this transformation, companies must start looking for their competitive advantage not just from the individual functions of the company but from its integrated functions.

Integration is the key. Functional barriers must be dismantled so that the individual excellence of each of the functions can be coupled to the overall excellence of the company. The success of the enterprise depends upon appropriately balancing the functions. Each of the functional areas needs to develop strategies that are complementary to the overall business plan, which in turn is complementary to the customer's preferential ranking of cost, quality, availability, and features. These functional strategies must be synthesized with finance and human resource strategies to compose the overall business strategy of the company.

To be successful, manufacturing companies must understand the series of linkages that compose the value chain of their business. Three value chains exist at HP and most other corporations. One is a conversion value chain that runs from procurement through manufacturing, distribution, and sales. Some industries add to that postsales support, vendor management, and the like. Another value chain is the design value chain that extends from the conception of a new product through its development and on into the manufacturing and conversion chain. A third is the marketing value chain, which begins with concepts and positioning and eventually culminates in sales strategy.

The organization must be managed not as a series of discrete functions but as a continuous process from procurement to sales and postsales support; from bright idea to successful introduction into the manufacturing flow; from identification of a market to the articulation of the sales message.

THE PROCESS OF CHANGE: STRETCH OBJECTIVES

There is, however, a fundamental problem in creating a climate for change within the organization. That problem is best expressed by a saying attributed to HP co-founder Dave Packard, "Tell me how a person's measured and I'll tell you how he behaves." The best intentions in the world will fail if the measurement and reward systems do not support the objectives. Setting goals will not create change unless top management follows through with appropriate incentives.

At Hewlett-Packard there have been significant changes in the internal measurement systems that can be traced back to the late 1970s. How was this change fomented? Primarily, top management established for the corporation a number of "stretch objectives" that demanded functional integration. The strategic vision underlying these stretch objectives was that HP needed a tighter coupling of the functions in order to achieve competitive advantage. Pursuit of these goals helped shatter, within HP, the myth of exclusivity. In its progress toward becoming a global, time-based competitor, HP focused first on a stretch objective for quality and more recently on one for product development and response time. The lesson from this experience is that, by focusing on quality and time, a firm can simultaneously reduce cost.

Stretch Objectives for Quality

In 1979, HP was a $1-billion corporation and was highly profitable. It held a dominant market share in instrument products and hoped to gain a dominant share in some other products. Yet despite its solid situation, the new chief executive officer, John Young, stunned his divisional managers by setting the first of several stretch objectives for the corporation. His challenge: improve the quality of HP's products by tenfold in the 10 years to 1990 (A $10\times$ quality improvement objective). Furthermore, Young proposed using warranty failure rates as the yardstick for measuring this objective. The general managers were shocked because then, as now, HP prided itself on the high quality of its products. The warranty failure rate of Hewlett-Packard products was among the lowest in the industry. Low cost and high quality were considered to be mutually exclusive. The sales and marketing teams had successfully pointed out to customers that the price of Hewlett-Packard equipment reflected its high quality. Some wags in the industry even claimed that HP stood for High Price. Clearly, many HP managers thought, focusing on quality would result in even higher prices for the products. It didn't seem to make sense to focus on quality when warranty costs claimed only about one percent of the corporation's costs.

The focus on improving quality came about after internal studies conducted in the late 1970s showed that fully 25 percent of HP's manufacturing cost was attributable to poor internal quality. One study examined the space usage in a large division and found that of the 90,000 square feet, over 48 percent was used for reworking products that did not work the first time, or for storing inventory, and only 15 percent of the space was used for real value-added activities. If everything went right the first time, most of the 48 percent nonvalue-adding space would be unnecessary.

Young's strategy worked: By 1985, HP engineers had achieved a three orders of magnitude improvement in its printed circuit board manufacturing, with first-pass failure rates down from over 1,000 to less than 30 parts per million. And by 1986 the company as a whole was well on its way to reaching the stretch objective set by Young. As the decade closed, HP's warranty failure rates per thousand dollars of shipments were fully tenfold lower than in 1979. And today, the quest for further improvements continues just as relentlessly.

At HP the challenge for the 1980s was quality improvement and not cost reduction, yet as a result of that focus on quality the overall manufacturing costs have been significantly reduced. Recent internal audits of the quality focus showed that for every $1 saved from warranty failure rates, over $5 was gained from other productivity improvements. These savings included over $500 million in inventory reductions and $250 million in accounts receivable reductions in addition to the $400 million savings in warranty costs. Whereas the quality focus led to cost reduction, it is clear that with HP's old understanding or methods

EXHIBIT 3
Commitment to Quality

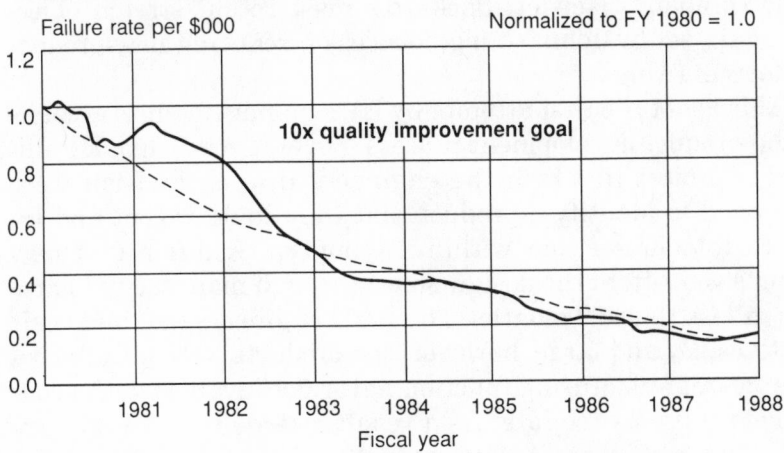

a stretch objective to reduce manufacturing cost would *not* have led to a commensurate increase in manufacturing quality. As many people have now realized, there is no trade-off between cost and quality. Focusing on quality—and doing it right the first time—leads to lower costs. The paradigm of exclusivity has been broken. (See Exhibit 3.)

BECOMING A TIME-BASED COMPETITOR IN NEW-PRODUCT DEVELOPMENT

Breaking the cost/quality exclusivity paradigm is not enough to stay competitive in the 1990s. It is also important to note that the leading Japanese firms, who have already achieved world-class quality, are now focusing on time-to-market. According to one commentator, Japan's central preoccupation is getting ideas into action quickly. This is the heart of time-based competitiveness and this is the competitive agenda for the 1990s. Hewlett-Packard introduces an average of one major new product a day and, historically, approximately two-thirds of the company's revenue has come from products introduced within the last four years. The new-product development cycle is of vital importance to the company. Hewlett-Packard's most recent stretch objective, again set by John Young, focuses on reducing new-product development time.

This is not the first attempt by HP's management to accelerate the product development process. Several years ago, HP initiated a project in which the engineers in each division were challenged to identify a product that they could invent and introduce into production within 12 months. A number of new products went from the design phase through manufacturing release in less than 12 months. The R&D engineers had met their objectives. By and large, however, the products were failures because the manufacturing function never got involved with products before it was too late. As a result a large number of engineering changes were necessary in the first six months of the product's life. The failure of these attempts to speed up the new-product development cycle was due to lack of functional integration.

The latest stretch objective tries to avoid the pitfall of functional ownership. The important metric, break even time (BET), is not the time from the idea until manufacturing release, but the time from idea until when the product returns its development investment through profits from sales.

Break Even Time is a very simple measure. Basically, it measures the R&D investment from the concept phase through the sales phase. Break Even Time is the time a project takes to recoup the R&D investment through sales revenues and profits. It requires that HP's functions work together in unison with a goal of reducing the number of engineering change orders. Hewlett-Packard's objective is to halve the time it takes to develop and recoup the development investment. The BET objective cannot be met unless R&D designs products that are innovative, easy to manufacture, and what the customers want.

In 1983, McKinsey reported findings regarding new-product development in industry (see Exhibit 4). McKinsey considered an industry in which there was 20 percent market growth, 12 percent annual price erosion, and a five-year product life, characteristics that closely resemble the industry in which HP com-

EXHIBIT 4
Factors Affecting Profit

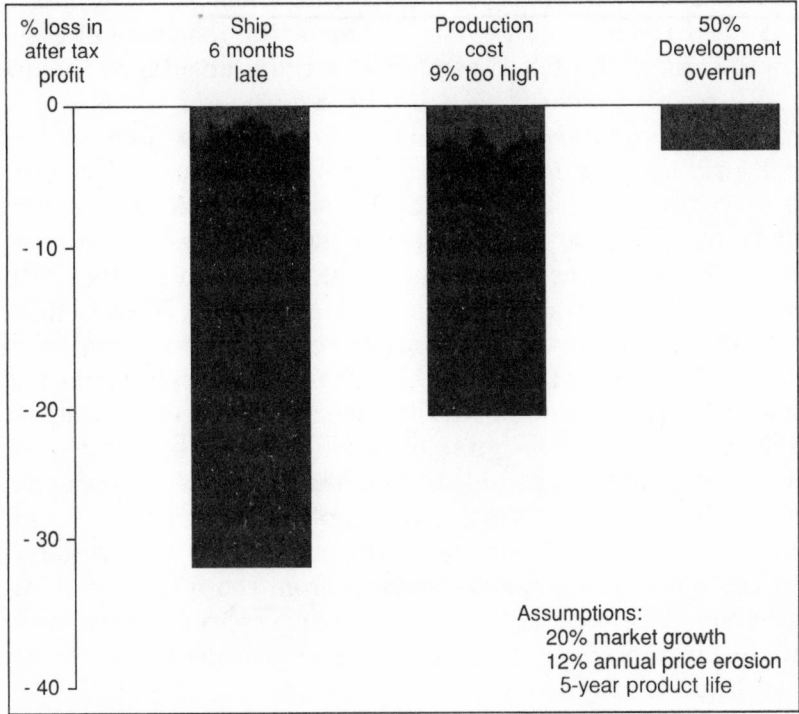

petes. The McKinsey model demonstrated that a 50 percent overrun in the development cost of a new product may result in a 3 to 4 percent decrease in profit. If production costs are 9 percent too high, a 20 to 22 percent decrease in profits results. Shipping that new product six months too late, however, results in a 32 percent reduction. Because of the new stretch objective, HP is now entering the next decade focused on reducing the elapsed time in the product-development cycle. Its divisions will understand why customers buy the products and will apply techniques such as total quality commitment, setup-time reduction, and the integration of information within the organization to achieve its BET objectives. HP once focused all of its time and effort on precluding cost overruns in product development, the area having the least impact on profitability. Now engineers at HP use detailed financial models in their project planning that clearly show the impact of various elements, including time, on the profitability of each product under development.

A need remains, however, to control the tendency of engineers to expand the list of components they consider in the design of any new product. (*As Mohan Kharbanda points out in Chapter 7, Xerox refers to this as the "clean sheet" approach to design—reinventing the wheel.*) It is this choice of components that drives the cost of the product throughout the value chain. Ford Motor Company recently conducted a study to determine which of its functions had the greatest impact on product cost. Ford found that the design function determined 70 percent of new-product cost. Hewlett-Packard once exercised the control of component selection through what was called "open lab stock." A large area was set up in R&D where the engineers could pick up the components to design their products. R&D was controlled only by manufacturing's ability to determine what components were put in the open lab stock. During the design phase, for instance, the engineer could decide to use any 20 ohm-resistor. However, they typically selected only from those readily available in open lab stock. By ensuring only preapproved parts were available in lab stock, manufacturing controlled the parts designed into the products.

The advent of computer-assisted design (CAD) has rapidly replaced the need for building early physical prototypes or "breadboards." Now design engineers are at liberty to develop products with any components that they can model electronically. The need for open lab stock has gone, and with it too has gone manufacturing's control over parts to be used by developers. Recognizing this, HP has spent considerable time and money to develop component product and process databases that provide the design team with critical information on approved parts and processes. Designers use assembly scoring and design-rule checking early in the process to determine whether or not the prospective product is manufacturable. HP also understands exactly which parts are preferred through linkages into its vendor and procurement systems. The impact of particular components on the product life cycle is factored into the selection process. With this information at hand, the engineers can model not only the performance characteristics of the system under development but can also model its cost, reliability, and manufacturability.

A specific example of a project that was affected by the new development time objectives occurred at HP's Roseville Terminal Division based in Roseville, California. In 1985, the management team decided to develop a computer terminal that cost less than $1,000 on a project that also included specific goals for time, quality, and features. History did not augur well for the success of this venture. HP had started many previous projects in the lab to develop a $1,000 terminal. Each time, however, the teams had missed the objective.

The reason for these past failures was a "disease" known as "creeping elegance" from which HP has suffered for most of its 50 years. Creeping elegance basically means that an engineer starts off with a great product idea, and the idea becomes even greater, and more costly, as it progresses through the design cycle. On this particular project at the Roseville Terminal Division, however, the project team surpassed its cost objectives. In fact, it came out with a family of terminals priced from $375. Not only did HP Roseville exceed the cost objective, it also produced a terminal that had an 85 percent quality improvement and that is available on very short lead times—all from a product made in America. HP has realized a healthy increase in profit margin on this new product.

Success when focusing on availability and quality has not been limited to Hewlett-Packard. In the pocket pager division of Motorola, an aggressive project was launched to reduce manufacturing time per product from days to minutes. In order to compete against its Far Eastern competitors, Motorola believed that although it could achieve parity in cost, quality, and features, the firm had the advantage of being based close to the market. Management decided to gain a competitive advantage through high availability. By building a totally automated factory in Florida, Motorola is now able to build and ship pagers to customers within hours of receiving the order. The real secret to Motorola's success was in focusing on the entire customer order-to-fulfillment process. An automated facility would have been worthless without the direct linkages into the order processing system. Motorola's objective was to improve order-entry-to-ship cycle times by over 85 percent. In doing so they found their defect rates halved, part counts decreased by 25 percent, and manufacturing efficiencies improved by more than 150 percent. Here again, the trade-off between cost and quality, cost and availability, and so on, has been proven false.

TIME COMPRESSION IN MANUFACTURING

Manufacturing must balance three primary components—the process, the resources, and the information. To meet quality and time-compression objectives, manufacturing has embraced the concepts of Total Quality Commitment (TQC), Just-In-Time (JIT), and Computer-Integrated Manufacturing (CIM). Together they have been instrumental at Hewlett-Packard in achieving aggressive stretch objectives and, therefore, warrant further explanation. Moreover, it should be recognized that these techniques can be applied in any function or supporting function of the corporation; they are not merely manufacturing tools.

The most important cultural change created by the 1980s quality stretch objective at Hewlett-Packard has been the institutionalization of TQC as a management practice at every level of the corporation. TQC is applied to every process—not just those in manufacturing, but also those in sales, administration, marketing, and R&D. A definition of TQC commonly used at Hewlett-Packard is shown below in Exhibit 5; most progressive companies have something very similar. TQC is not a program for the year but a fundamental philosophy that views the business as a series of processes that can all be improved through understanding (i.e., measuring them) and simplifying (i.e., eliminating the ones that are not necessary). TQC results in a singular focus on the process. A quality process, whether it be a manufacturing process, a sales process, or an accounting process, will yield a quality product. A quality process is one that is known, that can be measured, and that can be continually improved through understanding. Alternatively, if the focus is on the product alone it is unlikely that the same success will be

EXHIBIT 5

Total Quality Commitment
A *philosophy* whereby all elements of a business are viewed, by all those involved, as *processes,* and are capable of being *continuously improved* through understanding and simplification.

achieved. The essence, then, of TQC is being process driven, rather than event driven. There are many well-documented cases of TQC in action. At Hewlett-Packard one of the earliest case histories was documented at its Japanese subsidiary, Yokagowa Hewlett Packard (YHP). There the operators and the process engineers, together, reduced the defect rate of one key process from 4,000 parts per million to less than 10 parts per million in three years. This same HP subsidiary went on to become the first American-owned winner of the prestigious Deming quality prize.

Just-in-Time has been widely publicized over the past few years as many companies have adopted the techniques with almost missionary zeal. In fact, hundreds of firms have used a recent video produced by HP's Greeley, Colorado, plant to introduce employees to the basic principles of JIT in manufacturing. However, many still think that JIT is an inventory, specifically a work-in-process inventory, reduction technique. JIT, as practiced at HP and other leading companies, is the manifestation of an attitude that abhors waste. Waste might well be idle inventory, but it can equally be time wasted in nonproductive processes, such as setting up a machine for product changes. There has been much publicity on the impact of JIT here in America, and there are many examples of significant reductions in setup time. But those examples pale in comparison to the results reported by Cohen and Zysman. They reported that setup times in the "metal-abuse" industry have been slashed to *seconds* from the industry norms measured in hours and even days.[2] (See Ex-

---

[2]Stephen S. Cohen and John Zysman, *Manufacturing Matters* (New York: Basic Books, 1987).

hibit 6.) By focusing upon, and relentlessly reducing, the waste of setup time, a number of Japanese companies are rapidly moving to a position in which they can produce products in the same quantity and mix as the incoming order rate. This allows them to move to the ultimate goal of the time-based competitor—mass producing for a market of one, or in other words, enabling extreme customization of the product (features) while enjoying the cost and quality benefits of mass production.

A specific example at HP can be cited in the process of order insertion, the insertion of the electronic components into the circuit boards. In 1984, some leading HP divisions were taking approximately 24 or 25 minutes to change over from one product to another. That statistic has changed dramatically. At the Roseville division, for instance, they can produce a sequence of different terminals despite the many model variations. Each terminal is configured differently and can be produced at the run rate of the orders without interruption for setup or changeover. A bar code scanner reads the required configuration as the board passes into the auto-insert machine. The computers control the downloading of the correct sequence that ensures that the right parts are inserted into the board.

EXHIBIT 6
Setup Time Reduction

	Machine	Initial Setup Time (hours)	New Setup Time (minutes)	Implementation Time (years)
Toyota	Bolt maker	8	1	1
Mazda	Ring gear cutter	6.5	15	4
	Die casting machine	1.5	4	2
MHI	8-arbor boring machine	24	3	1
Yanmar	Aluminum die caster	2.1	8	2
	Cylinder block line	9.3	9	4
	Connecting rod line	2	9	4
	Crank shaft line	2	5	4

Source: Stephen S. Cohen and John Zysman, *Manufacturing Matters* (New York: Basic Books, 1987).

Hewlett-Packard and Automation

Although JIT has had a lot of press, it is exceeded in hyperbole by that which has been written about CIM (computer-integrated manufacturing). There are probably as many definitions of CIM as there are CIM consultants (and there are more than enough of those). A definition popular at HP is getting the right information to the right people at the right time to achieve their business objectives—a surprising definition because it mentions neither computers nor manufacturing. And that is why it is so powerful. CIM at HP is a process for linking new and existing technologies and people to manage cooperating activities across the manufacturing enterprise to achieve a competitive advantage. CIM, then, is the JIT analog for information resources.

Together, these three tools, TQC, JIT, and CIM, are a powerful combination. But they must be approached in the right order. The processes of the business must be understood (TQC) before waste elimination or simplification (JIT) can be effective. Similarly, the waste must be eliminated before information can be successfully automated (CIM). Failure to observe this basic tenet of "understand, simplify, and automate" will result in automating unnecessary complexity.

One very important lesson from the experiences of Hewlett-Packard and other companies has been that technology alone cannot integrate the business of the 1990s. As powerful as technology is, its effectiveness is limited by the firm's management methodology. If technology is applied to the current management practices (if the old is automated), then much of the positive benefits of technology will be dissipated. Technology must be seen as an enabling tool for changing the way the company of the future does business, much as the steam engine and the development of the railroads irreversibly altered the economic landscape of 19th-century America. Success, therefore, depends upon the highest levels of the corporation's management providing the environment for cultural change. This change cannot be achieved by following the traditional ways of doing business. It is up to the leaders of the corporation to set stretch objectives that clearly delineate common, measurable goals and then to monitor religiously the progress toward these goals.

The stretch objective has been fundamental in causing change at Hewlett-Packard, but HP is not the only company that has seen incredible competitive improvements by setting for itself breakthrough, or stretch, objectives. Motorola, one of the top electronics manufacturers in the United States, has a corporate stretch objective called the six-sigma challenge. The firm's goal is to reduce all defects in all processes to less than 3.4 parts per million.

Setting a breakthrough objective is a powerful management tool. However, it is unlikely that the objectives could be met without the underlying functions being equipped for the challenge. All of the stretch objectives that these corporations have labored under, from HP's $10 \times$ quality objective to Motorola's six-sigma challenge, have exhibited the following four characteristics. First, they have been driven by top management. Second, the stretch objectives have required a step function change in business practices, not incremental improvements. Third, they all have simple measurement systems that are simple to understand and serve to unify the various functions of the corporation. Fourth, the stretch objectives at Hewlett-Packard are integrative; they are not owned by any one function. Manufacturing, procurement, sales, marketing, and R&D all contributed to achieving the breakthrough in quality at HP. And they must all work together to achieve the breakeven-time objective. The breakthrough, or stretch, objective has been the catalyst in breaking the functional ownership of the strategic contributions (cost, quality, availability, and features) and, in turn, breaking the paradigm of exclusivity.

Much has to be done to ensure that America remains a competitive force in the world. The issues at stake are not ones of political dogma, or economic imperialism, but America's standard of living. Without a concerted effort to throw off the traditions of the past, to reform the corporations, Americans are destined to see their standard of living fall.

PART 3

TIME COMPRESSING LOGISTICS AND DISTRIBUTION

CHAPTER 9

TIME-BASED LOGISTICS

Robert Millen

Editor's Note: Robert Millen is professor of management in the College of Business Administration at Northeastern University. He has held faculty positions at Boston University, HEC-ISA in France, and UCLA. Professor Millen has published a number of articles on manufacturing strategy and material requirements planning. This chapter is based on a recent survey he conducted on JIT and time-based logistics.

INTRODUCTION

Many firms have implemented Just-In-Time (JIT) practices in their factories, and their successes have been well documented (see, for example, the discussion in Chapter 2). Improvements of substantial proportion in quality, productivity, space requirements, inventory levels, asset utilization, and response time have been achieved by companies in many industries.[1] Subsequently, these firms have included their vendors—both internal and external—in these improvement programs, and similar

---

[1]See, for example, R.W. Hall, *Zero Inventories* (Homewood, Ill.: Dow Jones-Irwin, 1983) and R. J. Schonberger, *Japanese Manufacturing Techniques: Nine Hidden Lessons in Simplicity* (New York: Free Press, 1982).

levels of effectiveness have been attained in their vendors' plants.[2]

As Toyota discovered, however, time saved in manufacturing can be dissipated in distribution. In 1982, Toyota found that the sales and distribution function was generating 20 to 30 percent of a car's cost to the customer—an amount that exceeded Toyota's cost to manufacture the car. Furthermore, while manufacturing could produce the car in 2 days, from 15 to 26 days were required to close the sale, transmit the order to the factory, get the order scheduled, and deliver the car to the customer. Toyota assigned its factory engineers and managers to its distribution function to determine how to improve these operations. This meant introducing into the sales and distribution function the same types of programs that had been found to be effective in reducing cost and time in the factory. Examples of these programs include statistical process control, single-unit processing of orders and information, and the removal of interfunctional boundaries.[3]

A recent study of the 500 largest manufacturing firms in the United States indicates that JIT programs are moving from the factory to the transportation function. Almost three-fourths of the respondents who reported JIT programs in their companies noted that the application of JIT has led to changes in the modes of carriage used for inbound and outbound movement.[4] As might be expected, rail was the biggest loser in both inbound and outbound movement due to the smaller and more frequent

---

[2]See, for example, A. Ansari and B. Modarress, "The Potential Benefits of Just-In-Time Purchasing for U.S. Manufacturing," *Production and Inventory Management*, Second Quarter 1987; D. Bartholomew, "The Vendor-Customer Relationship Today," *Production and Inventory Management*, Second Quarter 1984; C. Hahn et al., "Just-In-Time Production and Purchasing," *Journal of Purchasing and Materials Management*, Fall 1983; E. J. Hay, "Implementing JIT Purchasing Phase I: Top Management Commitment," *Production and Inventory Management*, January 1990; and C. O'Neal, "The Buyer-Seller Linkage in a JIT Environment," *Journal of Purchasing and Materials Managment*, Spring 1987.

[3]G. Stalk, "Time—The Next Source of Competitive Advantage," *Harvard Business Review*, July-August 1988.

[4]R. Lieb and R. Millen, "JIT and Corporate Transportation Requirements," *The Transportation Journal*, Spring 1988.

shipments caused by small-batch manufacturing. Aside from JIT manufacturing programs, other factors—including deregulation of the trucking industry—have caused the shift from rail to truck. This same survey, however, revealed that the five criteria in carrier selection, in order of importance, are on-time performance, responsiveness, tracing capability, extent of route network, and price. Given the effectiveness of truck versus rail based on these criteria, it is not difficult to understand why the revenue of Class One railroads in 1980 was twice that of long-haul trucking, whereas in 1985 the trucking industry reported revenues 11 percent higher than the rail industry.[5]

FASTER, BUT NOT NECESSARILY BETTER

Consider a pre-JIT factory in which quality is measured in parts per 100 defective; inventory turns are five per year; machines break down throughout the course of a day; an item is worked on for 3 percent of the time it is in the factory; and forklifts are used to move multiple pallet loads of parts at one time between successive operations scattered throughout the plant. Faced with this situation, some firms chose to automate: they installed the latest flexible manufacturing system (FMS), CAD/CAM, and robotics and tied them together with sophisticated, computerized information systems. Another approach adopted by a different group was to improve quality through employee training and the application of total quality control procedures, reduce setup times through single minute exchange of dies (SMED) programs, institute total preventative maintenance procedures, and re-layout the plant. Consideration of automation began only after the factory was functioning more effectively.

By now, we know which solution works best. The question is not whether there is a place for automation and computerized information systems but, rather, *in what order* these programs are implemented. Automating before eliminating waste and de-

---

[5]R. Roberts, "Railroads Team Up with Trucks in Order to Compete Better," *Modern Railroads*, February 1988.

veloping effective manual systems has provided minimal payoffs in the office.[6] Similarly, by 1987, General Motors had spent more than $40 billion on factory automation, but GM still is plagued with quality and productivity issues. In fact, one of the most productive plants in the GM system is its joint venture with Toyota in Fremont, California, which utilizes comparatively old technology.

Regrettably, "automate first" strategies also are being followed in distribution and transportation. To improve operations, some firms have shifted to air transport to speed physical flows and turned to electronic data interchange (EDI) to speed data flows. Other organizations have started by applying the basic steps from JIT manufacturing to these operations. Again, it is not a question of whether firms should utilize air transport and EDI, but whether these are the places to begin.

The choice between air and surface transport involves a trade-off between cost and time, rather than an option to improve cost and time performance. Viewing transportation in this way is equivalent to thinking that there is a trade-off between quality and cost or between service level and inventory size. Without fundamental improvements in distribution, substituting air for surface movement is equivalent to substituting a faster, more expensive machine for a slower, less expensive machine.

Consider the following example: The average time to get a load delivered intermodally is 17.8 days, in which half of this time is spent in terminals.[7] How much of the time and cost involved can be eliminated without a change to air transport or the introduction of EDI? Wal-Mart, in many instances, has cut out links in its distribution system through direct shipment to its retail stores; Sears has moved in the same direction.[8] A survey conducted by the consulting firm of A. T. Kearney reported

---

[6]G. Koretz, "Has High-Tech America Passed Its High-Water Mark?," *Business Week*, February 5, 1990.

[7]H. L. Richardson, "Focus on Quality," *Transportation and Distribution*, November 1989.

[8]D. G. deRoulet, "Doing More with Less," *Transportation and Distribution*, August 1989.

that costs for leading-edge logistics organizations average 4.9 versus 9.1 percent for all manufacturing firms. These leading firms also provide better service on order-cycle times, line-item fill rates, and on-time delivery. This survey further indicated that improvements must be progressive. In other words, firms fail when they try to make changes through system development all at once.[9]

Subsequent sections of this chapter address changes made by firms that seek first to improve their current operations and, then, to apply technologies to those operations. These changes have been applied successfully by carrier firms and private carriers. Not surprisingly, these change programs share common attributes with those applied successfully in the factory.

ATTENTION TO QUALITY PERFORMANCE

To improve on both time and cost dimensions, firms must examine their basic operating assumptions and decision rules regarding transportation. The first is the notion of quality or performance. On-time performance, which must be measured and tracked over time, is more critical for JIT programs than other programs. Measuring on-time performance, however, does not assist management in identifying what to do to improve this dimension. What else, then, should be measured? While some factors vary across settings, all systems should measure on a regular basis the following factors:

1. Percentage of damage and affected merchandise.
2. Maintenance records and procedures.
3. Driver certification levels and programs.
4. Equipment availability and cleanliness.
5. On-road safety record.
6. Response time to customer inquiry.
7. Transit performance—that is, time between points A and B, mean and variance.[10]

---

[9]Reported in *Traffic Management*, October 1989.
[10]See, for example, R. D. Frick, "Confronting Intermodal Challenges in a Global Mar-

These ideas may seem straightforward, but a 1982 survey conducted by the A.T. Kearney firm discovered that 85 percent of 1,000 shippers and carriers employed no productivity programs in physical distribution.[11] A later survey, released in May of 1989, of the readership of *Traffic Management* reported that 36 percent of responding shippers had carrier evaluation programs.[12] This finding indicates increased attention to performance during the years from 1982 to 1989. But, it also means that 64 percent of the respondent shippers still utilized no programs for carrier evaluation.

Most firms find it necessary to reduce the number of vendors—whether in manufacturing or distribution—in order to devote sufficient attention to their development.[13] Lieb and Millen[14] reported that 78 percent of the JIT firms in their study had reduced the number of carriers they used in prior days. For example, 3M reduced its number of transportation companies from 1,200 to 330.[15] GM has selected Ryder to provide most of the required distribution services for its Saturn Project. Ryder will dedicate 310 employees, 200 tractors, and 250 trailers to this GM contract.[16] Other examples of shipping firms embracing the "partnership" approach include Exxon and Eaton,[17] and another example of a carrier firm willing to do the same is St. Johnsbury.[18] These arrangements have led to a fundamental change in the relationship between carrier and shipper firms.[19]

---

ket," *The Private Carrier*, July 1988; W. E. Greenwood, "Strengthened Profitability, Service Quality, and Infrastructure to Trigger Intermodal Changes," *The Private Carrier*, July 1988; J-I-T, Inc, "What Makes a JIT Program Tick?," *Traffic Management*, February 1987; and A. Montgomery, "JIT: Methods and Practices," *Production and Inventory Management*, October 1986.

[11]C. G. Drugan, "Improving Productivity in Transportation," *Handling and Shipping Management*, September 1982.

[12]F. J. Quinn and J. A. Cooke, "Special Report—Transportation Quality," *Traffic Management*, May 1989.

[13]See Footnote 2.

[14]R. Lieb and R. Millen, "The Response of General Commodity Motor Carriers to Just-In-Time Manufacturing Programs," *The Transportation Journal*, forthcoming.

[15]F. J. Quinn and J. A. Cooke, *Traffic Management*, May 1989.

[16]J. D. Schulz, "Ryder Joins Partnership with Saturn for JIT Distribution," *Traffic World*, October 30, 1989.

[17]C. G. Drugan, *Handling and Shipping Management*, September 1982.

[18]B. Hull, "How to Make a Partnership Work," *Transportation and Distribution*, June 1989.

[19]R. Lieb and R. Millen, *The Transportation Journal*, Spring 1988; Ibid, forthcoming.

To improve distribution, however, firms must do more than simply reduce the number of carriers used or bring the function in-house. Union Carbide makes this case. The company's percentage of on-time pick-up has improved from 65 to 98.3 percent; its percentage of on-time delivery has improved from 85 to 98 percent. This success is due in part to the efforts made by Union Carbide to assist its external and internal carriers; for example, it provides videotapes on tank-inspection procedures.[20] But other programs for enhancing logistics quality and reliability are needed as well. Logistics consultant, Allen Ayers, notes that statistical process control in distribution has resulted in annualized savings of $50,000 to $250,000 in many applications.[21] The benefits of a total preventive maintenance (TPM) program in transportation include the ability to schedule as much as 80 percent of vehicle maintenance work, to have vehicles going *when* they are supposed to, to increase safety, and to provide greater fleet availability. One fleet manager reports that every dollar spent on TPM yields savings of $7 on repairs; further, TPM's return on investment is greater than $100,000.[22] Fuel usage is another area in which firms can improve effectiveness in logistics. One study indicates that 30 percent of all fuel usage is due to the operator. Hence, driver training is critical as equipment changes, as is periodic certification to ensure continued application of proper techniques.[23]

MOVING JUST-IN-TIME OUT OF THE FACTORY

Several programs that compose the essential parts of implementing JIT in factory operations can be adapted to a firm's distribution and transportation operations. These programs can be

---

[20]F. J. Quinn and J. A. Cooke, *Traffic Management*, May 1989.

[21]L. H. Harrington, "Stop Service Problems Before They Start," *Traffic Management*, October 1989.

[22]M.E. MacDonald, "Why Your Fleet Needs an Ounce of Prevention," *Traffic Management*, November 1989.

[23]L.H. Harrington, "How to Maximize Fuel Efficiency," *Traffic Management*, February 1989.

implemented whether a firm provides its own transportation or uses a common carrier. Instituting a partnership-type arrangement can provide the motivation for a common carrier to focus on firm-specific issues; otherwise, the carrier may not perceive the benefits of customizing its service for one firm's requirements. The manager of a contract carrier firm involved in this kind of a partnership arrangement stated, "We are an extension of the shipper's company; we just get paid differently."[24] The remainder of this chapter describes the counterparts of JIT programs for transportation and distribution and the impact their implementation has had on performance.

Reduced Setup Times and Costs

In manufacturing, there is a basic trade-off between the cost of inventory and the cost of a setup. That is, as more units are produced, the setup cost decreases because it is spread over more units. At the same time, however, the cost of carrying inventory increases as more units are produced. A similar relationship holds in transportation. Sending a truck fully loaded or only partially loaded results in nearly the same cost; the driver's wages and the lost capacity are independent of the quantity shipped. Hence, a full shipment leads to a reduced cost per item in terms of the transportation charges each unit must absorb. Total inventory, on the other hand, will increase because a truck load quantity usually is more than the amount needed on a daily or more frequent basis.

Transportation customers prefer using local vendors when possible.[25] In some cases, firms require their suppliers to locate within a 100-mile radius of the customer's plant. This approach has been implemented by Toyota in Japan and by most U.S. automobile manufacturers. Given America's extensive and effective interstate road system, however, overnight delivery can be provided within distances up to 500 miles.

Related to the movement of suppliers to closer locations is the introduction of smaller vehicles. These vehicles may be dedi-

---

[24]R. Lieb and R. Millen, *The Transportation Journal*, forthcoming.
[25]R. W. Hall, *Zero Inventories*.

cated vans that supply daily requirements from a vendor to a customer or that collect parts from multiple vendors, which then will be consolidated for shipment in large trailers or double trailers to the customer.[26] A variation on this theme is that two firms, located near to one another, can share space on a truck that brings shipments from different vendors to each firm. Magnetics Peripherals, Inc. uses this approach with shipments coming from Los Angeles to Oklahoma City.[27] A further variation is to pick up parts from multiple vendors, ship them to a consolidation warehouse, form a mixed load, and then ship to the user.

All these approaches still present the backhaul problem. The setup cost of a carrier can be the equivalent of two setups; the first, going from vendors to customers; the second, returning from the customer to the vendors. Honda Motor Company of America, which receives container shipments from Japan, has addressed this backhaul problem. Honda procures U.S. products to export to Japan to reduce the company's shipping costs. For example, Honda has shipped live cattle to Japan in aircraft containers.[28]

Vendors can implement still other ways to reduce the time involved in transportation. Similar to factory setup-reduction programs, carrier firms can observe loading and unloading operations, eliminate unnecessary operations, and perform off-line preparation activities. The impact of such time-reduction efforts is evident in Toyota's increased number of deliveries of cars per day to U.S. dealers. Daily deliveries by each of its trailers have increased from two to five.[29]

Reduction in Material Movement

To reduce handling, many firms have altered the layout of their shipping and receiving departments. At the Wrangler division of VF Corporation in Greensboro, North Carolina, denim is de-

---

[26]"Ford Motor Awards JIT Contract," *Production and Inventory Management*, May 1987.

[27]J. R. Carter et al., *International Sourcing for Manufacturing Operations*, Monograph No. 3, Operations Management Association, September 1988.

[28]R. D. Frick, *The Private Carrier*, July 1988.

[29]"Toyota Using Own Trucks in US," *Traffic World*, November 13, 1989.

livered to the location where it is used in the factory. Many other firms have adopted this approach as well.[30] This approach requires top-quality certification of the vendor and the transport firm. Moreover, the plant must have or create unloading docks located at the various points of use in the shop layout. Process flow diagrams can be employed effectively to determine new layouts that will reduce movement and speed handling.[31]

Another approach used by some firms is to dedicate specific pieces of equipment and personnel to specific accounts. In situations in which requirements are regular and recurring, this leads to a thorough understanding of the customers' actual requirements. This approach, however, raises the issue of utilization of resources. Because most firms receive and/or ship products between 7 A.M. and 4 P.M., carrier equipment that is dedicated to a single account may be idle during times other than the period needed to perform preventive maintenance. Some trucking firms thus have instituted two-driver teams so that vehicles can be operated during the interim hours while in route to a distant customer.

Cross-Training and Expanded Operator Responsibility

Most machine operations require the human operator only to load and unload the machine; the human operator does not perform the actual machining operations. The use of U-shaped or S-shaped lines enables individual shop workers to operate multiple machines. Workers, furthermore, are trained to perform multiple tasks so that lines can be rebalanced in light of changes in demand.

In transportation, however, one driver cannot operate more than one vehicle at a time. But there are ways to extend the

---

[30]See, for example, G. C. Jackson, "J-I-T Production: Implications for Logistics Managers," *Journal of Business Logistics* 4, no. 2 (1983); H. Miller "J-I-T: Some Textile Industries Call It 'Linkage'," *Purchasing*, April 9, 1987; and A. Montgomery, *Production and Inventory Management*, October 1986.

[31]G. Gagnon, "Use Process Flow Diagrams," *Transportation and Distribution*, November 1989.

tasks performed by the driver. Flexible job assignments, for example, can require drivers to assist in offloading their vehicles rather than waiting for someone else to offload. In another instance, drivers now create the freight bills after a pickup.[32] Several transportation firms require their drivers to call in after each pickup or delivery if they are experiencing difficulties (e.g., the vendor's shipment is not ready, traffic is unusually heavy, or traffic is slowed due to an accident). A few firms even have added on-board cellular telephones for this purpose. When these problems occur, therefore, contingency planning can begin immediately.

Homogenized Scheduling

Homogenized scheduling is the practice of mixing the sequence of products manufactured. In many firms, the schedule for making four different products is to make product A for three days, product B for three days, product C for three days, and product D for three days. Homogenized scheduling calls for mixing the products such that the production sequence might look like: A B C D, A B C D. To do this, however, requires reduced setups, flow-type layouts, and so forth.

Some trucking firms have adapted to homogenized production by developing "milk runs" whereby one truck visits several vendors on a daily basis before delivering to the customer's plant.[33] Instead of each of five vendors sending a truckload per week to supply the customer's weekly needs, five truckloads per week are still shipped, but each truckload contains only the daily requirement from each vendor. Likewise, some vendor firms now provide similar delivery: Rather than shipping truckload quantities to each customer site at specified intervals, one vehicle delivers small quantities to multiple customers each day. General Motors calls this kind of vendor operation "peddling."[34]

---

[32]See J-I-T, Inc., *Traffic Management*, February 1987.
[33]*Production and Inventory Management*, May 1987.
[34]D. Blumfield et al., "Reducing Logistics Costs at General Motors," *The Private Carrier*, April 1987.

Other steps are necessary to enable transportation to support this new production schedule. "Milk runs," as noted, can be employed to collect parts from different vendors on a regular basis. The parts then must be placed on the trailer in the order in which they will be used. Some firms, instead of loading materials from front to back, have installed partitions in the trailers so that materials are loaded side by side and removed as their delivery requires. Alternatively, materials can be loaded from front to back if the vehicle is canvas-sided. Another alternative is to load the heaviest items first along the bottom of the vehicle and add successive pickups on top of them. In all cases, the purpose of the onloading system is to make it possible to offload a "product's worth" of parts in the proper order for each customer. Saab currently uses this approach with its vendor that provides automobile seats. The seats are shipped every two hours in the sequence in which they will be used in the assembly plant.[35] Finally, a transportation schedule requires precise pickup and delivery times. A delivery time of "sometime next week" will no longer suffice; today's delivery time must be, "we will be there next Tuesday at 10 A.M. (plus or minus 15 minutes)."

Kanban Techniques

The Kanban systems that many firms have implemented in transportation generally take the form of empty containers. The number of empty containers dictates the number of parts to be provided in the next delivery. Using containers helps ensure that the firm will perform an operation in the right way. In shipping, this may entail the use of standardized containers because many JIT firms employ standard, reusable containers so that material arrives ready to use.[36] Standardized containers aid in determining if the correct numbers are present, but they also provide several other additional advantages. These containers

---

[35]S. Wandel and R. Hellberg, "Transport Consequences of New Logistics Technologies," Working Paper, IIASA, Laxenberg, Austria.

[36]H. L. Richardson, "Cutting Packaging Also Cuts Costs," *Transportation and Distribution*, October 1989.

reduce weight and disposal costs, generate less wasted space in trailers, and require less labor to pack, unpack, and prepare. Several companies now use "garment on hanger" containers, so that garments move from the container to the retailer's shelf without the unpacking, repressing, and hanging costs incurred when garments come packed in regular boxes.[37]

Quality Control Circles

Quality control circles are perhaps the best known element of the JIT system. Transportation firms have used this element in their own operations to examine how to perform maintenance more effectively, how to load and unload more rapidly, and how to improve driver effectiveness.

Just-in-Time manufacturing firms need to develop circles that include the vendor, the shipping firm, and the customer. Together such groups can address issues that intersect among them, such as the loading sequences of parts and the timing of receipts and deliveries. At least one carrier, for example, sends its drivers on sales calls when considering a partnership agreement. In another example, two plants routinely sent one truckload of parts per day from each site to the other. The QC circle, which included both drivers, determined that the drivers, rather than staying overnight near their destination sites, should switch trucks at midpoint and return to their home plants with the inbound parts.

Subsequent Steps: New Technologies

Upon implementing the foregoing programs, effective opportunities for automation, air transport, and the application of other technologies should be apparent and appropriate. These opportunities include the RailRoader Mark V, a rail and road trailer with a simple conversion,[38] or, spine rail cars, which can

---

[37]D. A. Clancy, "Perishability of Garments Calls for Value Added Service," *Transportation and Distribution*, October 1989.

[38]"RoadRailer Goes European," *Transportation and Distribution*, May 1989.

carry different types of cargo and switch types rapidly. Several computer applications also can lead to enhanced capabilities for distribution and transportation operations. For example, Nabisco uses a PC-based claims processing system through which it has reduced its outstanding claim balance from over $2 million to less than $250,000, a figure that is still dropping.[39] The Ore-Ida Company has installed automated order consolidation and carrier selection software producing a savings of nearly $700,000 per year.[40] Air Products and Chemicals, Inc. uses a computerized routing procedure that has led to a 6 to 10 percent reduction in operating costs, estimated to represent at least $1.5 million per year.[41]

Telecommunications technology also offers opportunities for improved performance. These include satellite communication systems, bar code scanners and transmitters, and EDI. Currently, the 100 largest truck carriers and all Class One railroad companies have EDI capability. Many shippers require EDI capability of their carriers.[42] EDI links can be employed for bills and invoices, bills of lading, and shipment status reports. For many firms, obtaining status reports requires EDI links among the shipper, customer, and carrier firm.[43] This system also enables customers to determine the status of their shipments more easily. Northern Telecom, for example, dials into Consolidated Freightway's computer to track shipments, pickups, material in-transit, and the day's deliveries.[44] Vendors in the textile industry send shipping information to customers so they can prepare patterns for cutting material in advance.[45] Furthermore, EDI can be used for international movements to clear customs prior to actual landing of the goods in the United States.

---

[39]F. J. Quinn, "Profile of a Partnership," *Traffic Management*, February 1989.

[40]C. Ebeling, "Preparing for Spud Shipments Proves Profitable," *Transportation and Distribution*, October 1989.

[41]W. Bell et al., "Improving the Distribution of Industrial Gases with an On-line Computerized Routing and Scheduling Optimizer," *Interfaces*, December 1983.

[42]R. Lieb and R. Millen, *The Transportation Journal*, Spring 1988.

[43]P. A. Trunick, "Distribution Information Management Gains Importance," *Transportation and Distribution*, October 1989.

[44]J-I-T, Inc., *Traffic Management*, February 1987.

[45]H. Miller, *Purchasing*, April 9, 1987.

THE FUTURE: VALUE-ADDED WAREHOUSING

Implementation of JIT in both the factory and transportation systems frees up space in current warehouses. As supply chains shrink in time and in the number of links, opportunities for adding value to products at the warehouse become evident. To reduce delays caused by customs procedures, for example, Kawasaki, USA has a foreign trade zone (FTZ) on its factory site in Nebraska. A delay of up to five days to clear customs at West Coast ports is eliminated because most FTZs are operated on an audit basis with fewer other distractions.[46] Other examples of processing in the field include adding labels for private branding of consumer products such as batteries and canned vegetables. Furniture can be shipped in component pieces for subsequent assembly at the warehouse or at the customer's site.[47]

CONCLUSION

All of the programs discussed in this chapter are designed to provide greater value to the customer in a shorter period of time and at a lower cost. Many of these programs can be implemented within the existing resources of a firm. Implementation, however, requires rethinking past ways of doing things, mostly small things—but whose cumulative impact provides strategic advantages. Although pressure abounds to improve performance on many dimensions simultaneously, the first step is to break away from the mindset of cost-versus-time or reliability-versus-cost. One critical task is to determine how a firm's transportation and distribution can provide a competitive advantage.

---

[46]J. R. Carter et al., *International Sourcing for Manufacturing Operations.*
[47]K. B. Ackerman, "Value-Added Warehousing Cuts Inventory Costs," *Transportation and Distribution*, July 1989.

CHAPTER 10

THE DISTRIBUTION REVOLUTION: TIME FLIES AT FEDERAL EXPRESS

Frederick W. Smith

Editor's Note: In this chapter, Frederick W. Smith, the CEO of Federal Express Corporation, provides key insights into how his firm pioneered, and now dominates, the U.S. air express industry. He explains how Federal Express was guided throughout by a time-focused strategy. Smith also describes his vision of the future development of distribution systems, based on Just-In-Time concepts and time-based delivery. He argues that to be competitive in tomorrow's global marketplace, a firm must become a time-based competitor. This point is made emphatically:

There's no doubt in today's business milieu, "time" is a key factor in determining which companies grow and succeed and, conversely, which ones stumble and fall[1]

---

[1]All italicized passages are excerpts from a keynote address by Frederick W. Smith, Chief Executive Officer of the Federal Express Corporation, to the "Time-Based Competition Conference." The conference was held at the Owen Graduate School of Management, Vanderbilt University, (Nashville, TN) on December 18, 1988. The address was entitled "Time: Tomorrow's Weapon for Global Competition." Federal Express statistics have been updated to reflect conditions in Spring, 1990. Thomas Plath and Joseph Blackburn also assisted in writing this chapter.

TIME-BASED COMPETITION IMPACTS DISTRIBUTION AND LOGISTICS

Business strategists are realizing that philosophical changes are necessary to remain competitive in today's marketplace. Those philosophical changes result from the recognition that competing is becoming increasingly dependent on response time. Time-based competition (TBC) is more than just another catch-up program targeted at narrowing the margin between U.S. and foreign competition. It is a strategic way of thinking, a mindset that can change the way companies do business.

According to Business Week, *"The new math of productivity points to time as a manufacturer's most precious commodity."*

. . . Time as both a commodity and a competitive weapon is an emerging issue that business people can't ignore if they expect to survive in this increasingly competitive world.

Now, of course, time as a control tool is not new—witness time-and-motion studies. But time as a strategic tool is a new concept, and those who find ways to shave a month, a week, or even a few minutes off their production cycle or distribution channels while simultaneously improving quality and service will have a decisive edge over their competitors.

In fact, the marketplace will see the demise of marginal firms who don't adopt time-based strategies. And the longer they wait, the faster they will fall. In short, where everything else is equal, time-based strategies become a key factor in widening the gap between those who adopt them and those who don't.

The intensified movement toward time-based competition will usher in a revolutionary business paradigm comparable to the introduction of mass production at the turn of the century.

Time-Based Competition Brings Change

Time-based competition has two driving forces—the Just-in-Time strategy and the belief that effectiveness is contingent upon closeness to the customer. In a TBC setting, management desires to compress the throughput time in manufacturing. This compression is achieved primarily by reducing setup times, producing small batch sizes, and working in tandem with suppliers. When time is compressed in manufacturing, an incentive is also created for management to take time out of distribution, and thereby to deliver the product to the customer more quickly. A historical perspective of time-based competition provides insights into its strategic advantages.

To understand the beginning of the emphasis on time compression, one needs to look back to Japan, shortly after World War II. With the help of the American business strategist W. Edwards Deming, the Japanese began to assemble and mobilize their productive resources. And since they were in a catch-up mode, Japanese manufacturers placed emphasis on efficiency and the elimination of waste, and eventually, they reduced production-run time. In the process, they raised productivity to new heights. The rest, of course, is history and is well documented.

Many Japanese companies also began to see that traditional, multilayered distribution systems were inherently inefficient and that they impeded both production and product quality. Moreover, they soon discarded the traditional view that quality is separate from production. Instead they began to integrate quality into the production process itself through a monitoring and adjustment system known as statistical process control.

The goal of the system was to obtain optimum productivity from the minimum amount of material, labor, and equipment. The rationale behind downsizing was to make each employee responsible for the quality of every part he handled. This meant that each person had to examine the material passed to him before he could operate his machine, thus establishing a perfect quality feedback loop.

Japanese companies learned, too, that there was no real reason to stock large quantities of raw material. The basis for their reasoning was that with quality built into the process and little or no rework required, the stockpiling of material—with its attendant need for space, recordkeeping, and capital—was redundant, time-consuming, and costly.

The solution was obvious: deliver the necessary raw materials or components only as they were needed. One important aspect of this new thinking was to change from a system of production where work in progress is "pushed" from workstation to workstation to one where it is "pulled" from station to station. Thus, production proceeds on the basis of demand scheduling as opposed to the traditional forecast or "push" scheduling. But to do so required a novel way of thinking about production and its relationship with quality and distribution. And, of course, the resulting manufacturing system has become known as "Just-in-Time" or JIT.

However, JIT attracted only limited attention outside Japan until the recession of the early 1980s. That downturn forced manufacturers in other countries to reexamine every aspect of production and distribution and to find ways to cut costs, shorten the production cycle, and speed up the process of new-product development from concept to market.

Thus, as a result of severe domestic economic problems coupled with foreign competition, many American manufacturers are now trying to regain their competitiveness in the world market. Savvy manufacturers are examining every aspect of their operations with a view toward shortening each time interval that occurs within the product or delivery cycle.

WHAT MAKES THE CHANGE POSSIBLE?

Time-based competition is indeed changing distribution and logistics; however, other rudimentary changes needed to occur first. These changes methodically infiltrated the business world and set the stage for the onset of TBC.

"The end of this century will be marked by fundamental industrial change," forecasts Jim Barksdale, chief operating officer of Federal Express. "In the same way that railroads changed industry at the turn of the last century, jet transportation will completely change distribution."[2] Businesses do not have to wait until the end of the century to validate Barksdale's forecast.

Improvements in transportation have already contributed to the success of the TBC philosophy. Highway transportation

---

[2]"FedEx: America's Warehouse," *InformationWEEK*, May 16, 1988, p. 28.

has been upgraded through improvements and additions to the interstate system. Fuel-efficient trucks allow companies to cut costs and increase expenditures on additional time-saving measures. Barksdale's forecast iterates the need for companies to study the positive effects that time compression can have on their logistics and distribution systems. Historically, the benefits of time compression have been proven repeatedly.

Of course, the notion that time is a distinct business component is not totally revolutionary. The history of trade shows that businessmen have always been cognizant of a time factor in the movement of goods.

Throughout history, whenever a faster mode of transport has been made available, it has eventually become the primary distribution mode. Consider that trucks supplanted the rails as the quickest and most efficient means of distribution, just as railroads had once replaced water-borne transport.

Now the jet airplane has become the primary method for shipping high value-added goods. And with continued technological developments in avionics and new research in hypersonic craft, this method of transport will shorten the "time" continuum even further.

The fact is that quite a few companies in the United States and Europe saw this trend. They not only paid attention to the revolution in logistics systems that Japan pioneered, but they also adopted certain aspects of the new logistics sciences and even developed newer versions to meet the special needs of their marketplace.

The transformation of transportation is one of the integral factors in the changes which TBC brings to the business world. A second influential factor is the boom in information and telecommunication technology. Traditional methods of distribution and logistics are beginning to appear archaic in light of current technological innovations. For instance, instantaneous tracking now allows distributors to pinpoint the location of any shipment from the time it is picked up to the time it arrives at its destination. This tracking capability also can be provided directly to the customer via computerized telecommunication channels. This instantaneous tracking is a singular example of an innovation that supports the two driving forces of TBC, namely, JIT and getting closer to the customer.

Interestingly, the story of Federal Express's creation and its growth mirrors the evolution of what has been termed the "distribution revolution." The two together parallel the emergence of a "world economy." And, all three are linked by cause and effect.

As early as the mid-sixties, it was clear that advances in automation, mechanization, and computerization would one day require radically different approaches to distribution. Indeed, it became evident that traditional freight and postal services would not be able to satisfy what was seen as a growing demand for a new standard of service that emphasized expeditious, time-handling.

Federal Express was created in response to this need and, in the process, pioneered the modern U.S. air express industry, which has become a $10 billion industry.

Federal Express can help customers control their inventories by providing both operational and informational systems designed to serve their very specific needs.

For example, by maintaining a positive tracking system, combined with terminals placed in customers' own offices, inventories can be positively controlled within the Federal system as easily as they would be if stored in the customer's warehouse.

Moreover, Federal Express should be thought of as being more than just a means for moving time-sensitive documents and parcels. Rather, FedEx transport jets should be thought of as warehouses moving at speeds up to 500 miles per hour. The cargo is secure and is virtually assured of reaching its destination on time. Every time. Around the world.

Absolutely, positively. That is the claim that Federal Express makes and lives up to each and every business day. The remarkable growth of FedEx during its short existence is a case study in time-based competition.

THE FEDERAL EXPRESS STORY

Federal Express was born of humble beginnings. As a Yale undergraduate in 1965, Frederick W. Smith was time-pressed to complete yet another term paper. In the term paper, for which he received just a C, Smith maintained that the air freight industry's future was by no means assured and that the passenger route systems used by most air freight shippers were totally wrong for freight distribution. Smith's hastily completed paper contained no details of a system design of an airline network with technical and administrative support systems. Rather, the theme of the paper was simply that there was a huge market for the efficient movement of small, high-priority shipments. This theme provided the basis for the purple, orange, and white operation now known as Federal Express.

Fred Smith's father had amassed a fortune in founding and leading Dixie Greyhound Bus Lines. From this fortune came the $8 million inheritance that Smith used to create his overnight delivery phenomenon. In 1971 and 1972, Smith secured venture capital to purchase and lease 33 aircraft and establish the business at a total cost of $56.1 million.

Despite a very slow start, Federal Express slowly built credibility among the corporate ranks. The company was able to do this only by offering low-cost, high-quality service to its customers. The company's slogan of, Absolutely, Positively Overnight, provided the company throughout the years with the motivation necessary to be the time-based competitor it was supposed to be. The company historically differentiated between itself and its competition by marketing itself as the mover of the most vital and time-sensitive parcels and documents.

Smith accepts the management principle that structure should follow strategy. Major changes were made in the company's organization chart during the first decade. Smith feared complacency and wanted to be sure that the company's structure would facilitate the organizational plan. He still fears complacency and continues to implement strategies that will maximize effectiveness and efficiency.[3]

In 1990, after only 17 years, FedEx is a $7-billion-plus company traded publicly on the NYSE. It is facing strong competition both domestically and internationally, not only from UPS, DHL, Emery, and others, but also from the onset of facsimile

---

[3]Robert Sifafoos, *Absolutely Positively Overnight* (Memphis, Tenn.: St. Luke's Press, 1983).

usage to communicate time-sensitive documents. The firm's competitive strategy is now focused on the international delivery marketplace. Two-day delivery globally is its goal for the 1990s. As it pursues this global goal and that of maintaining dominance of its niche in the U.S. market, FedEx will continue to base its competitiveness on its ability to deliver on time.

Early on, Federal Express recognized the need for Just-in-Time inventory procurement on the production side and rapid, random, and time-certain distribution of products and spare parts on the sales side. Federal Express evolved as a unique, integrated transportation system designed specifically to meet those needs. As a result, it emerged as a major transportation company in a relatively short period of time.

From the beginning of its express operations in 1973 . . . Federal Express now delivers over one million shipments a day, with shipping access to 99 percent of the U.S. population and to 126 countries worldwide . . . employs over 86,000 men and women around the world . . . has a fleet of nearly 400 cargo aircraft— either on hand or on order—including 58 jumbo McDonnell Douglas DC-10 and MD-11 freighters—the largest widebody freighter fleet in the world . . . making daily flights to Europe, Asia, Australia, and Latin America.

Federal Express has developed a telecommunications and computer system that is one of the largest on-line networks in the world with over 80,000 computer terminals of one type or another. The system is so sophisticated that it can track any shipment anywhere in the world within 30 minutes of a request . . . records who signed for the shipment and the exact time it was delivered. And all this works efficiently and profitably. . . . Revenues for the 1990 fiscal year totalled nearly $7.0 billion.

Federal Express has been successful in large part because of its belief that time is a manageable business component. And the firm practiced that belief. . . . The key to the future success of any modern business enterprise lies in its ability to manage its use of time throughout the product/service delivery cycle.

Federal Express is recognized as the world's largest full-service, all-cargo airline. The company has seen phenomenal growth since its humble beginnings in 1973. In contrast to the handful of packages processed on that first night of service, the 86,000 employees of FedEx are now responsible for processing an average of 1.5 million shipments daily. The shipments are processed at two major hubs, Memphis and Indianapolis, and are routed to locations throughout the world. FedEx offers direct service to 99 percent of the U.S. population. With the acquisition of the Flying Tiger fleet, FedEx nows serves 303 airports and more than 125 countries worldwide. FedEx has proven that the "hub-and-spoke" delivery system is an efficient operational strategy. Many airlines and most other distribution companies recognized this system's advantages and implemented similar hub-and-spoke operations.

FACTORS IN FEDERAL EXPRESS'S SUCCESS

The Links between FedEx and JIT

Federal Express can easily be considered a benchmark in the utilization of time-based competition. In its distribution operations, FedEx strives for a synchronous flow that is characteristic of the best JIT processes. Daily operations are so masterfully scheduled that they are reduced to clockwork. With an accent on speed and teamwork, the workers minimize wasted motion. A day with the workers of Federal Express in its Chicago facility exemplifies the scheduling genius to which so many customers and employees have become accustomed.

Orders are called in to Federal Express all day long during every business day of the year. Many of these orders are received in the morning and can be scheduled for the afternoon run. Others are received later in the day and are communicated to couriers via CRT terminals in the trucks. Still others are left in drop boxes at various locations throughout the Chicago area.

By 1:30 P.M., the couriers leave the garage area along carefully plotted routes which will minimize driving time. The drivers pick up packages until late afternoon, struggling to maintain their schedules despite the rush hour traffic. They arrive back in the garage by 5:30 P.M. to assist with the early sorting of packages. The couriers later pick up those parcels left in drop boxes throughout the city. The principal sorting of packages gets underway by 8:30 P.M., and couriers hurry back to assist, knowing that the 10:15 P.M. deadline will approach much more quickly with fewer workers on location to sort packages.

Everyone at the facility assists in the sort after 9:30 P.M., knowing that the loaded vans must be on their way to the airport at 10:15 P.M. At least 50 couriers, cargo handlers, and supervisors help to transfer parcels from the conveyer belts to cargo containers. Other employees prepare plastic bags for Courier-Pak envelopes and the Overnight Letters. Still others tenderly prepare the hazardous shipments for loading. All workers have a sense of mission and frequently encourage one another. Two high school girls scan the bar codes on the Courier-Pak and Overnight Letter envelopes; this process enables tracking of the packages from this locale until they reach their final destinations. At 10:05 P.M., there is a final push for completion; by 10:10 P.M., the final containers are sealed. The huge van moves out of the garage at precisely 10:15 P.M. headed for the Kennedy Expressway and O'Hare airport.

The van arrives at the airport at 11:20 P.M. along with a number of other vans from Federal Express facilities in the Chicago area and throughout the Upper Midwest. By 11:40 P.M., the vans have been unloaded, and the FedEx DC-10 is on its way to Memphis, the company's primary national hub. After a routine flight, the jet arrives at Memphis International at 12:38 A.M. Other planes from Boston, Newark, and Los Angeles have already arrived. Before the barrage of planes ends at 1 A.M., approximately 80 purple, orange, and white DC-10s and 727s have arrived in Memphis.

Cargo from planes is unloaded and sorted by 5,000 employees. As the items move steadily down the conveyer belt, the ZIP codes are keypunched into computer terminals. This process allows an automatic routing of the packages to the specific area of the hub where the cargo containers will be loaded. About 800,000 packages and documents are processed between midnight and 3 A.M. in Memphis.

By 3 A.M., all planes have been refueled, checked mechanically, and reloaded. Federal Express planes start departing Memphis International at 2:50 A.M. and continue departing one per minute for the next hour. The Chicago-bound plane is slotted to take off at 3:44 A.M. and does so, on time, as usual. The nightly miracle has occurred once again—a three-hour flurry of activity that effectively and efficiently routes hundreds of thousands of time-sensitive parcels from almost every corner of the United States to almost every corner of the United States.

The return flight to Chicago taxies in to O'Hare at 5:15 A.M. The cargo van, waiting nearby, is loaded efficiently and leaves for the downtown facility within minutes. The van pulls into the station at 6:20. Workers arrive a few minutes later at 6:45 and sort the cargo by routes. They have just one hour and 35 minutes to complete the sorting and coordinate their route plans for the day ahead. The work is completed by 8:15 A.M.

Promptly at 8:20 A.M., the vans leave to make deliveries in downtown Chicago. The couriers must deliver every Federal Express package by 10:30 A.M., and they do so approximately 99% of the time. Snowstorms, traffic jams, and mistakes account for the remaining 1 percent. The promise to deliver all packages by 10:30 A.M., of course, is a key feature of Federal Express' timely service. To accomplish this, the morning sorting process at the distribution center must be done accurately and efficiently. Packages must be sorted quickly and allocated to the correct truck routings. Incorrect assignments cause delays and late deliveries.

Federal Express attacked this problem in the same way that world-class manufacturing firms have attacked production problems with JIT. In a recent analysis of the Natick, Massachusetts, distribution center near Boston, time-compression techniques were used to speed up the morning package sort.[4] The first step in the process was an operations, or value-added, analysis, just as one would do in reducing a setup time in a JIT process. According to one of the managers, "We suspected there was downtime built into our process. But we didn't know just where it was until we completed our data gathering. We used the data to redesign our sorting system and put our resources where they could do the most good."[5] By charting the process and isolating wasted steps, the FedEx team was able to use resources

---

[4]George H. Labovitz, "Speed on the Cycle Helps Companies Win the Race," *The Wall Street Journal*, October 30, 1989, p. A10.

[5]Ibid.

more effectively to speed up the sorting operation. As a result, vans now leave 15 minutes earlier. Customer satisfaction has increased along with the percentage of on-time deliveries.

At Federal Express, time-based competition extends to claims and customer complaints. In his speech to the Vanderbilt Time-Based Competition Conference, FedEx CEO Fred Smith explained his shock when he learned that Federal Express averaged about one month to settle claims from customers for damaged and late shipments. Although claims were a miniscule fraction of the millions of packages handled, Smith's reaction was absolute and positive: as a time-based competitor, FedEx should handle claims the way it handles packages—overnight. Soon, Federal Express was handling over 95 percent of its customers' claims within 24 hours.

A day with the Federal Express crew is an exercise in precision. All the drivers, loaders, pilots, and other employees know that time is FedEx's competitive advantage. All know that without the advantage, they are, as their ad says, "Just like everybody else." Federal Express has mastered the TBC concept and has used it to maintain an enviable advantage.

USE OF INFORMATION TECHNOLOGY

Two interrelated factors involving modern information processing have contributed to the success of Federal Express. Behind each of these factors is FedEx's adherence to the premise that time is money. "Good companies are becoming aware that the management of time is just as important as the management of money," says Barksdale. "They are able to react more quickly and capitalize on change."[6] Ron Ponder, senior vice-president for information and telecommunication systems for Federal Express, agrees: "A lot of us feel that the concepts of JIT and time-based competition coincide directly with the direction of our company."[7]

---

[6]"FedEx:America's Warehouse," *InformationWEEK*, May 16, 1988, p. 29.
[7]Ibid.

One factor behind FedEx's success is its dedication to total customer service. Thirty call centers handle an average of 297,000 calls each day. FedEx will ship any package up to 150 pounds by 10:30 A.M. the following business day. From the time that a courier picks up the package to the time it is delivered at its destination, customers have access to the exact location of the package. This control is made possible through the company's newest package and document tracking and tracing system, the Cosmos IIB.

"Everyone must understand that Cosmos IIB . . . is leading edge technology that's going to differentiate our product and services from our competition," says Ponder.[8] Cosmos IIB makes use of the SuperTracker, a hand-held computer that scans, records, and transmits airbill and billing data directly to a customer service center after it has been downloaded into the DADS (digitally assisted dispatch system) unit in the van. "The state-of-the-art SuperTracker will assist with the same-day trace guarantees and the 30-minute response guarantee," says Dave Dietzel, Cosmos IIB project manager.[9] The development of the Cosmos IIB accentuates the fact that Federal Express is totally dedicated to timely customer service.

A second success factor in the Federal Express story is its operational efficiency. The Cosmos IIB is certainly designed to increase customer satisfaction. The use of the SuperTracker, however, is also a very efficient way of accomplishing a difficult, but necessary, operational task: tracking each of the 1.2 million packages processed daily. "The SuperTracker is a great device," said Allan McArtor, senior vice president of air operations. "It is clearly an ingenious application for operations, and it is going to revolutionize operations such as ours."[10]

---

[8]Gale Jones Carson, "Cosmos IIB to Differentiate FedEx from Competitors," *Update*, January-February 1987, p. 6.

[9]Ibid.

[10]Ibid.

FEDEX'S CHANGING ROLE: FROM DISTRIBUTION TO INVENTORY MANAGEMENT

Indeed, the management of time will be a major factor in the future strength of the world economy, a strength that will only be derived from the levels of efficiency and quality made possible by time management.

Any business that ignores this marketplace dictum will almost certainly find that its customers have gone elsewhere. Perhaps more importantly, any government that ignores this new trend will find that the standard of living of its people will stagnate or even decline.

This is not some far-fetched "someday" scenario . . . In terms of both domestic and international trade, most businesses are already moving toward a different kind of commerce—one that is based on components of logistics—service and time—in an attempt to keep pace with the inexorable forward movement of the marketplace.

Really, they have no choice.

Lee Iacocca thinks it is true of the automotive industry. In an interview that appeared in Fortune *magazine in 1988, the Chrysler chairman said, "A company with the best distribution system and the best service will win all the marbles . . ."*

Phil Schaecher, vice president of operations for Lands' End— a company whose net sales grew from $72 million in 1982 to $265 million in 1987—agrees that time-conscious logistics systems are critical. He said, "The era of allowing six to eight days for delivery is gone forever. People won't stand for it."

Robert Earley, vice-president of distribution for Williams-Sonoma—a company with a sustained annual growth rate of 20 to 30 percent—put it another way. He said, "People want instant gratification."

And to underscore that point, consider the experience of R.E.B., a catalog distributor who found that ground service delivery caused a high cancellation and return rate; after converting to air express, the cancellation rate was reduced by 70 percent.

Numerous other examples underscore the notion that time is a valuable and manageable business component. Think about *what McDonald's did for hamburgers. And think about the economic reward to McDonald's for understanding and applying time-based competition tactics—in 1987, the average annual sales per McDonald's franchise was $1.5 million.*

Federal Express has translated its vaunted operational efficiency into another strategic opportunity: expertise in inventory management. This expertise is now shared with companies such as IBM and Williams-Sonoma, who route their products through FedEx distribution channels. "We are a large distribution facility for IBM," says Chris Demos, one of the managers at FedEx most directly responsible for the success of the Cosmos system. "They would not be able to do these things for themselves."[11] Big customers, such as IBM, have FedEx terminals in their offices and can access the much-heralded Cosmos package-tracking system. These terminals allow FedEx to supply "logistics management" to a company, and, at the same time, move deeper into the company by taking over its distribution operations. In effect, FedEx is an IBM warehouse. "With the right control, why can't an airplane be a warehouse?" asks Chris Ponder.[12]

Williams-Sonoma is a San Francisco-based supplier of kitchenware, home furnishings and accessories, and lawn and garden goods. The company has posted remarkable growth recently: it averaged over 40 percent a year in sales growth for the past five years. The firm's five-year plan is to grow at least 25 percent a year. Why has Williams-Sonoma grown so quickly, and why is the company able to anticipate continued sales expansion? Williams-Sonoma located its distribution center in Memphis, thousands of miles away from its base of operations in California. One reason for the selection of Memphis was its centralized location. The primary reason, however, was to gain proximity to the Federal Express hub. Williams-Sonoma, as a time-based competitor, strives to be closer to the customer. FedEx gives the company a way to obtain superior, centralized inventory control and yet be only an overnight shipment away from the entire United States. Historically, getting closer to the

---

[11]"FedEx: America's Warehouse," *InformationWEEK*, May 16, 1988.
[12]Ibid, p. 29.

customer meant a network of decentralized distribution centers. Maintaining such distribution centers drove up inventory costs and was unwieldy to manage. Centralized inventories are easier to manage and more effective. With the new information and telecommunication technology, it is not necessary to have order processing and distribution in the same location. Williams-Sonoma can be headquartered in California (if they wish), but inventory can be held in Memphis and integrated into FedEx's overnight distribution machine.

Williams-Sonoma is already one of Federal Express's biggest customers out of Memphis, particularly during the holiday season. Williams-Sonoma also is able to take advantage of the extremely reliable scheduling process at Federal Express. Because of the proximity of the FedEx hub, the firm's distribution center can get a package to the hub later in the day than companies in other cities and still be certain that the package will be to its destination the following morning.

"In the mail-order business, the whole distribution angle is the product," says Richard M. Metzler, managing director of U.S. marketing for Federal Express. "If something isn't in the right place at the right time, the person trying to sell through that type of distribution is out of business."[13] Williams-Sonoma is far from out of business; it rang up over $220 million in gross sales for 1989. Williams-Sonoma has experienced the cost benefits of a time-based strategy by getting "closer" to their customers while retaining the operating efficiency of centralized inventory control.

Another problem in the context of distribution is the cost of obsolescence. With the proliferation of technology-intensive equipment such as computers and computer parts, medical equipment, chemicals and pharmaceuticals, and countless others, obsolescence is a real risk that creates significant costs. However, with rapid, time-certain air transport, manufacturers can skip traditional distribution channels, thus saving money and reducing the risk of obsolescence.

---

[13]David Yawn, "Catalog Businesses Choose Memphis for Distribution Hub," *Memphis Business Journal*, January 8, 1990, p. 33.

Federal Express realizes that customer satisfaction and operational efficiency are contingent upon time compression. FedEx is methodically overtaking each niche of the distribution market because of its demonstrated capability to provide timely service to each and every customer, whether that customer is an individual, a growing business like Williams-Sonoma, or an industrial giant like IBM. The firm is well on the way to its intended goal of becoming "America's Warehouse."

THE ROLE OF FEDERAL EXPRESS IN A GLOBAL ECONOMY

Although precise figures would vary by the type of business, there is ample evidence from the Japanese experience and from Federal Express customers of just how valuable JIT and reduced inventories can be. And the savings flow directly to the bottom line in the form of increased profitability and cash flow.

All the cost advantages of JIT/time-based manufacturing need not be cited, but a few points will illustrate. In the traditional "push production" system, raw material is ordered for the warehouse and is processed whether there is a demand or not. In such cases, it sits on a shelf until needed. The production floor is usually filled with work in process and rejects that are idle, waiting for disposition. It has been estimated that materials are actively in actual production for only 1 to 3 percent of the time they are in the factory. The remainder of the time, those materials are idle, work-in-process, rejects, or misplaced.

As long as material is sitting on a shelf, no value is being added. While there, it runs the risk of being damaged or becoming obsolete. And it is a drag on cash flow.

Think of carrying costs, which represent the cost of the space, capital, inventory, service, taxes, insurance. When material is sitting on a shelf or on the floor, those costs keep on adding up.

On the other hand, if there is only the absolute minimum inventory—ideally zero—turn rates increase, resulting in greater net cash flow; warehouse labor is reduced or eliminated; a more orderly flow of materials produces better customer service; and the fixed investment is reduced or eliminated, freeing capital for use in other areas. It is not uncommon for businesses converting to JIT to increase their turn rate by a factor of 10. According to one large manufacturing company, after converting a unit to JIT, the firm reduced inventory from over three months' supply in 1980 to just over three weeks' by 1986.

And consider the industry Federal Express represents—the air express industry—an industry that demonstrates the realities of time-based competition. FedEx is part of an industry that places a premium on service and timely distribution. Accordingly, it is a necessity to constantly analyze the existing distribution system, seeking always to upgrade its capabilities.

Federal Express is committed to developing new technologies and more sophisticated delivery and information systems. This commitment has led to a capital spending program of almost $1.5 billion in 1987–88 alone.

Federal Express must use new technologies because it has to stay on top of things. FedEx has and will continue to update because there simply is no other choice.

In the air express transport sector, the ability to meet time-based demands has begun already to alter significantly the views of an increasing number of business people regarding inventory. Owners and managers already burdened by the costs of production and delivery are beginning to understand that they can all but eliminate inventory overhead through time-certain delivery of goods.

The air express industry is destined to play a pivotal role in the emerging global economy and will be a strategic player—a critical link, if you will—in the time-based competitive environment.

Indeed, this area is the world. And the arena in which these time-based strategies will operate extends far beyond the confines of any one country or region. In fact, the playing field will encompass the globe. And like it or not, every person is part of the world economy.

A quote from a speech by MIT economist Lester Thurow illustrates the point:

> *You live in a world economy and you will continue living in a world economy the rest of your lives, even if you never leave your*

hometown—because you're competing with foreign producers, you're going to be selling to customers who have foreign competitors, and you're buying from foreign producers.

The global market will be won by the companies that can "get there firstest with the mostest."

The United States possesses an economy based on efficiency, one that not only relies on the coefficients of logistics, service, and time, but also blends them together into a formula for sound economic success.

As has been illustrated, there are numerous competitive advantages to using express transport. Just-in-Time adds value to the product, increases overall efficiency, and offers new opportunities for efficient organization. It forges even closer links between seller and customers.

Technological changes in transportation system capabilities are creating an era in which economic opportunities abound. But the potential for economic growth and further innovation is likely to be realized only in more open domestic and international markets than currently exist.

In spite of the willingness and ability of the air express industry to respond positively and firmly to the demands of time-based competition, the marketplace sometimes just cannot be won. Like other companies, Federal competes in many markets that are far from totally free. Although the potential is recognized for both national and global gains through a sophisticated air express operation, outdated international aviation treaties continue to pervade significant parts of the competitive playing field.

On such a playing field, it is increasingly difficult—and in some places well-nigh impossible—to meet the demands of time-based competition. Federal Express intends to lobby for economic and political progress—better trade agreements and fewer regulations—to keep pace with the realities and challenges of time-based strategies both here and abroad.

Again, there is no alternative. The profundity of approaching time as a manageable business component cannot be overstated. And what is out there is a global market. The changes in communication and transport which have taken place in just the past two decades have seen to that. The scheduled evolution of the Euro-

pean Common Market into an unfettered regional alignment by 1992, and the dismantling of trade barriers between Canada and the United States that will take place by 1999 in accordance with the new Free Trade Agreement, give further impetus and truth to this judgment.

As regulatory relaxation occurs, rapid, time-certain logistics systems are bound to increase in importance. And countries and industries that want to participate in this global market of the 21st century would be well advised to facilitate the type of transportation policies and activities that open the field to all players.

Surely, the dictates of time-based competition can tolerate nothing less.

CHAPTER 11

THE QUICK-RESPONSE MOVEMENT IN THE APPAREL INDUSTRY: A CASE STUDY IN TIME-COMPRESSING SUPPLY CHAINS

Joseph D. Blackburn

Time-based competition, under the name "quick response" (QR), is revolutionizing the textile and apparel industry in the United States. The movement has been spearheaded by large textile producers, such as Milliken and Company and du Pont, and major retailers, including J. C. Penney, K mart, and Wal-Mart, and has spread throughout the industry. The QR movement is an excellent model for time-compressing supply chains in other industries because of its emphasis on cooperative efforts among all members of the chain linking raw materials to the customer.

As in other industries, the catalyst for a QR movement was competition from offshore. In 1988, 50 percent of the apparel and textiles purchased in the United States were imported from the Orient. Drowning in a flood of imports, the domestic textile pro-

---

*Much of the material in this chapter is based on presentations to the "Time-Based Competition Conference" at Vanderbilt University, December 1988, by textile industry executives and on supporting data supplied by the consulting firm of Kurt Salmon Associates.

ducers realized that the future of their business depended on a healthy domestic apparel business.

Industry leaders, seeking a competitive advantage, perceived that nationalistic slogans would not be their salvation. For today's sophisticated, worldly consumer, waving the flag is not enough: an apparel producer might as well wave a white flag of surrender. An appeal to the emotions and patriotism of the consumer is less likely to succeed than one based on time, quality, and price. In other words, few consumers make a clothing purchase based on whether Old Glory is sewn onto the label. To win back the consumer, U.S. manufacturers had to find a more potent hook: an advantage based on being closer to the consumer than distant, offshore competitors.

The outcome of the debate over this issue within an apparel industry group called the Crafted with Pride in the U.S.A. Council was the formation of the QR movement. The thinking of the Council was as follows: There must be a competitive advantage based on our proximity to customers. We ought to be able to seize that advantage and to satisfy U.S. customers' needs more quickly than firms overseas.

The Council also recognized that closer to the customer means time more than distance. The name of the game in apparel is fashion, and in the fashion game timing is everything. Today's style is tomorrow's markdown. Successful firms are those that can ride the latest wave of fashion. They do not have to be on the leading edge, but they do have to be right behind the leaders—fashions tend to be set by the high-fashion houses, the couturiers in Paris, Milan, New York, and Tokyo. For the rest of the apparel manufacturers, response time is the key: quickly judge what is selling and get it through the pipeline and onto the racks before your competitors do. Be nimble and be quick. Next month or next season is too late.

The fact that most of the firms in the apparel industry with a time advantage were foreign obviously rankled the members of the Council. The Italian firm, Benetton, was highly publicized for its prowess in cutting the lead time required to move knitted goods from factories in Italy to retail stores in the United States. Using speed, Benetton capitalized instantly on shifts in consumer preferences and rushed the hot new colors and styles to their U.S. stores. In a market as trendy and ephemeral as teenage fashions, Benetton's quickness fueled a money machine. From the other side of the world, knock-off shops in Asia, working from FAXed sketches of the latest fashions, rushed new garments onto racks in U.S. stores faster than domestic suppliers. The U.S.-based apparel industry was allowing its sole natural advantage, time, to slip from its grasp. Retaliation came in the form of quick response.

Quick Response Defined

According to Andy Vecchione, director of quick response for Milliken and Company, "Quick Response is Marketing 101. That is, getting the right product at the right place, at the right time and at the right price."[1] The QR movement is a renewed effort to make that happen on the part of U.S. apparel producers. Quick response is a consumer-driven strategy. It substitutes a pull system for the traditional push system. The objective is to pull things through the pipeline and let the consumers decide what they want instead of the old-fashioned way of trying to forecast what the consumer wants and then push it through the pipeline.

THE LONG PIPELINE

Converting QR from a dream to reality required facts and figures. Where was the proof that there was value to being closer to the customer or to getting apparel on the rack faster? This was the charge given to Kurt Salmon Associates (KSA), a consulting firm with a knowledge base in apparel built on years of experience with clients in the industry.

The initial study that launched QR was commissioned by the Crafted with Pride in the U.S.A. Council in 1986. The simple objective of the study undertaken by KSA was to characterize the total apparel pipeline from raw materials into textiles, to

---

[1]Remarks by Andy Vecchione, "Time-Based Competition Conference," Vanderbilt University, Nashville, TN, December 1988.

EXHIBIT 1
The Apparel Pipeline

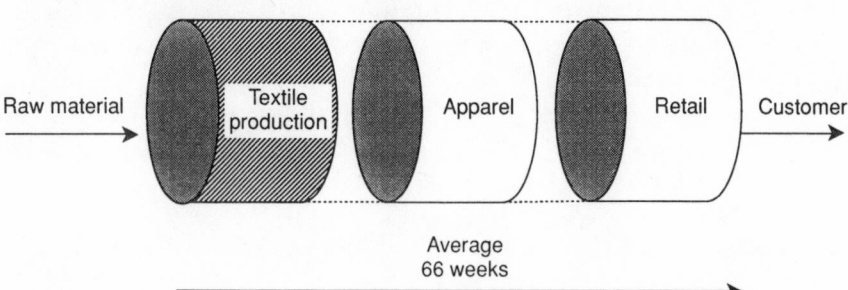

Raw material → Textile production | Apparel | Retail → Customer

Average
66 weeks

apparel, to retail and the consumer. How long was the pipeline? How much time was required at each phase?

Kurt Salmon traced the pipeline and discovered that it averaged about 66 weeks to turn raw material into textiles, to turn textiles into apparel, and to distribute to retail and to the consumer (see Exhibit 1). In a $125-billion per year industry, they observed a $25-billion efficiency loss. The majority of that efficiency loss, over $16 billion, occurred in retailing, and the major reasons were forced markdowns, out-of-stock items, and the cost of carrying excess inventory.

It is easy to see that forecasting problems plague an industry beset by a product pipeline that extends for more than a year. With that time horizon, the manufacturer can merely guess what the consumer will want by the time the product reaches market. The 1987 miniskirt fiasco is an excellent example of what can happen. Despite the enthusiastic support of the male population, the skirts flopped and a lot of money was left at the end of the pipeline in unsold skirts. Quick response provides a remedy by compressing the total pipeline, thus eliminating some of those efficiency losses by moving manufacturing closer to the consumer's buying decision.

WHY QUICK RESPONSE WORKS

The motivation for QR is derived from a fundamental principle of forecasting: forecast error diminishes in proportion to the time until the event. The forecast of tomorrow's weather tends to

EXHIBIT 2
Effect of Lead Time on Retailer's Stocking Decision

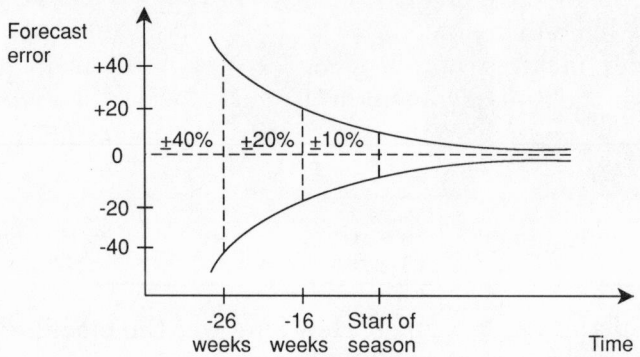

be more accurate than the prediction for a week from tomorrow. In apparel, the event is the consumer's purchase and the forecast interval is the time lag between the retailer's stocking decision and the consumer's buying decision. Taking time out of the pipeline moves the stocking decision closer to the customer in time and, therefore, decreases the level of uncertainty in the decision.

Exhibit 2 demonstrates the effect of a QR program on the retailer's stocking decision. In this scenario the buyer, under the conventional supply system for basic apparel, must order 26 weeks in advance of the start of the season. Under QR, cycle-time reductions make it possible to delay commitment until 16 weeks before the season (times would be reduced for fashion or seasonal apparel). In both cases the buyer is faced with uncertainty; however, the level of uncertainty increases as one moves back in time from the consumer's instant of purchase. Forecast error grows more than linearly as the time to forecast increases, forming a familiar shape known to forecasters as the "trumpet of doom." In this example, the forecast error increases from + or − 20 to 40 percent. Averaged over thousands of items, the ability to sharpen forecasts can cut markdowns and lost sales by millions of dollars.

A similar "trumpet of doom" could be drawn to portray the

effect of QR on merchandise replenishment decisions. Under QR, merchandise profiles are reviewed and orders made on a daily, rather than weekly or monthly, basis. Then, weeks are removed from the replenishment cycle by compressing order processing, order picking, and shipping cycles. With fashion items, the longer the replenishment decision can be delayed, the easier it will be for the buyer to project sales trends and forecast final sales.

Retailing in the United States has been growing at an average annual rate, adjusted for inflation, of 1 to 3 percent. The Wal-Marts and Limiteds are exceptions to the rule, but the average is under 3 percent annually. The competition is vigorous and there are malls everywhere you go in the United States today. So the growth rate is not due to a lack of stores or shopping opportunities. According to a textile industry executive, the problem is "sameness." That is, the same items are available to the consumer in several stores. As the executive explains, "You can go into any mall in the United States today and find at least two stores within that mall carrying the same basic item you want. So what happens as a consumer? A consumer goes into the store she normally patronizes and if she cannot find the style, the size, and the color she wants, she doesn't say anything, she just leaves because she knows she can buy that same item in at least one other store in the mall. So what happens? As a retailer you lose a sale, maybe you lose that slack sale that she was looking for and that's a great loss. You also lose the complementary shirt sale or the tie, the blouse, the belt, or other accessory that goes with it. That is also a great loss. However, the greatest loss is consumer loyalty, which jeopardizes the stream of potential future sales from that customer. Consumers across the United States today are pressed for time, particularly in families with both adults working. Those consumers will tend to shop the stores where they know they can walk in and find exactly what they are looking for."[2]

---

[2]Remarks by textile industry representative, "Time-Based Competition Conference," Vanderbilt University, Nashville, TN, December 1988.

The failure to respond quickly to consumer apparel preferences is one of the reasons why the mail-order catalog industry, the L. L. Beans and Lands' Ends, have done so well over the last few years. Actually, mail order is a misnomer for this industry; phone or FAX-order catalog industry is more appropriate. You can phone and always know on-line, in real time, if they have the item you want and that it will be shipped to you within 24 to 48 hours.

Pilot Studies

Recognizing that the lengthy pipeline made the U.S. apparel industry vulnerable, the Council commissioned several pilot studies to see if the pipeline could be reduced and to evaluate, with specific retail/apparel/textile partnerships, the advantages of faster response time. In each of these experiments, consultants focused on a specific product line as a control group, took steps to shrink the reaction time in the chain for the product, and measured the sales and profit potential from quick response.

The results of several of these pilot studies, carried out by KSA in the mid-1980s, are summarized below.

Producr	Quick Response Chain (Retail/Apparel/Textile)	Results
Jackets and leisure slacks	Belks/Haggar	Sales up 25%; gross profit up 25%; inventory turns up 67%
Tailored clothing	J.C. Penney/Oxford/Burlington	Sales up 59%; inventory turns up 90%; forecast error cut by more than half
Basic slacks	Wal-Mart/Seminole/Milliken	Sales up 31%; inventory turns up 30%

In the case of the Belks pilot study, analysts made a telling point by calculating that, to produce identical profits at retail, a non-quick response supplier would have needed a 28 percent lower wholesale price. This was an important message that the backers of QR wanted to send to decision makers in retail: To gauge profit potential, look at the total cost of the chain. Time compression is a powerful competitive weapon.

Over 100 of these pilot studies were conducted in the mid-1980s to launch QR. These studies were effective in getting the message across to the apparel industry that quicker response was possible through a coordinated, partnership effort. Time compression in the supply chain has a powerful dual effect on profits: Sales can increase while inventories are decreased. Revenues go up and costs go down.

THE SECRET TO QUICK RESPONSE: COMPRESSING TIME FROM TWO DIRECTIONS

Conventional approaches to time reduction in manufacturing are one-dimensional: The objective is to move product faster through the pipeline; that is, speed is attained by moving forward faster. However, the apparel supply chain, shown in Exhibit 3, is not atypical of other supply chains. In any processing chain converting raw materials to consumer products, there is flow in two directions: Product flows forward in value-added stages and information flows backward from consumers as demand is translated into replenishment orders.

The significant conceptual breakthrough made in the QR movement was recognizing that the supply chain is a system of bidirectional flows. Product flows forward from textile producer

EXHIBIT 3
Two-Way Flows in Apparel Chain

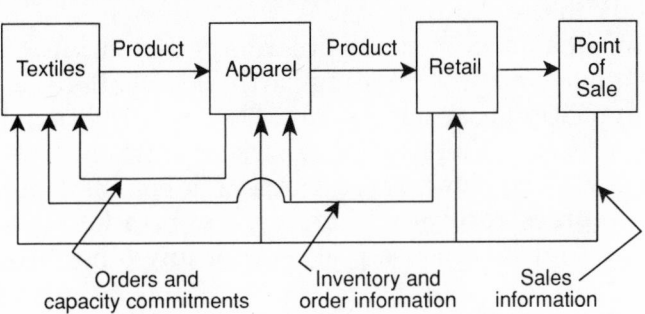

to the customer and information about consumer demand, orders, and capacity commitments flows backward to apparel and textile manufacturers. In control systems terminology, this is a system with product fed forward and information feedback.

The innovative feature of QR is that actions are taken to increase the velocity of flows in both directions. The QR campaign involves steps to move apparel forward faster; that is, to shrink cycle times at each stage of manufacture and the shipment times between stages. However, the unique features were those taken to improve the responsiveness of the system through improved, faster communication of consumer preferences back to all members of the chain. In subsequent sections we will examine the key actions taken to speed the flow of apparel and information.

PARTNERSHIPS—THE KEY TO SHARED INFORMATION

Quick response is built upon a partnership involving each link in the chain from the producer of textiles on through to the retailer who sells the garment to the customer. The desired partnerships in QR extend beyond the paired relationships between supplier and buyer that are found in JIT systems; each link in the chain shares information about sales, orders, and inventories with the others. Retailers communicate information about sales, not only to the apparel manufacturer, but also back to the textile producer. Mutual cooperation among all partners is required if the system is to succeed in its goal of increasing sales with less inventory in the total system. A successful partnership means higher inventory turns and improved return on investment for each link in the chain.

The first step in a quick response program is to find willing partners. As in other industries, finding partners is challenging because, as an industry spokesman states, "In your traditional relationships, what's out there today is lack of trust. Nobody trusts anybody. Breaking down those barriers is not easy, but that is the key to quick response. There's got to be a lot more communication. In quick response programs or any other program where change must be made, it is harder than the old way of doing business. A lot harder."[3]

Solid partnerships are a prerequisite for quick response. To shrink the pipeline, information must be shared and that can happen only when there is trust. Adversarial relationships block communication and inhibit the unfettered flow of information needed for effective supply chains, whether in apparel or in the JIT systems of other industries. All of the successful pilot programs in quick response were built around partnerships (for example, Wal-Mart/Seminole/Milliken; J. C. Penney/Oxford/Burlington), and these cooperative relationships generated improved profitability for each link in the chain.

Information Flows and Electronic Data Interchange

Established partnerships are the medium that cultivates a faster, freer information flow: the second key element in QR. Quick response depends on a timely and accurate flow of information to all the members in the partnership. Retail provides it to apparel manufacturing, who shares it with the textile manufacturer. Instant information about consumer preference shifts is needed throughout the chain to improve the system's reaction time.

In any distribution chain, apparel included, the traditional method of communicating demand information to the members of the chain is through a sequence of replenishment orders. When a sale occurs at retail, no one else in the chain is cognizant of it. The apparel manufacturer learns that a product is selling well at retail only when orders begin to flow back from the stores' buyers. Instead of instantaneous transfer, information about customer demand is collected and batched at the retail level until stocks have been depleted to the reorder point. Further back in the chain, the textile manufacturer learns about customer demand through orders for fabric from the apparel producer. Thus textile production is twice removed from the cus-

---

[3]Remarks by a textile industry representative, "Time-Based Competition Conference," December 1988.

tomer, and the batching of information into orders creates a long time lapse between the instant of sale and knowledge of the sale at the fabric production stage. Batching demand information into orders injects lengthy, dysfunctional time delays into the information transmission process between stages. Effectively, batch transfer of information moves suppliers further away from the customer.

New technology—specifically, barcoding, scanning, and electronic data interchange—is the driving force behind the improved information flow and data sharing needed to make quick response work. K mart, for example, has invested over $500 million in systems for QR and point-of-sale (POS) technology. Barcoding of each item is an essential part of QR and is becoming common in the apparel industry. Barcoding automates the tracking of each item at the stockkeeping unit (SKU) level at every stage in the chain, but most importantly at point-of-sale. Using barcoding to capture, electronically, consumer demand at time of purchase, the data interchange technology allows the retailer to do more than just keep accurate inventory and sales records. Information about the sale is immediately transferred back through the chain to the apparel manufacturer and the textile manufacturer. With an emerging industry standard for electronic data interchange (EDI), a true information partnership is technologically feasible; an item can be tracked at the SKU level at every point in the chain from raw material to customer.

Industry analysts point out that goods should be preticketed by the manufacturer and that the barcode should be affixed at that point. The best place to code is at the starting point, at manufacture. If the item goes uncoded to the retailer, two weeks of time can be consumed to get it through the distribution center. In addition, coding is more expensive at the retail level: 15 cents versus less than a nickel to code at manufacture. Anything that will shorten the cycle and get it on the selling floor faster is preferred. Industry standards are crucial to the spread of QR because the barcode is a license plate that each member of the partnership uses for identification and tracking.

A textile industry expert describes how the system works today: "The initial stage of involvement is to capture the point-of-sale data at the retail store. It is very important that we capture it at the SKU item. Traditionally, they have captured it at the style item level but, today, we must capture style, color, and size because it does not help you if you have a size 14 neck and they've got tons of 15s in stock. It has got to be the right style, size, and color that you want. That information has got to be accurate, it's got to be timely, and it's got to be shared with all the partners. The reason for that is one size doesn't fit all. Everybody has a different size. How many of you go out and buy a shirt today and it has 15 1/2 inch collar but a 33/34 sleeve? That means it only fits the occasional customer with a 33 1/2 inch sleeve."[4]

With point-of-sale data from barcoding and scanning, a JIT replenishment system becomes possible. Some retailers have established a model stock inventory; that is, a distribution of colors and sizes that they need to offer a correct presentation in the store. Every time they sell a garment, they replace one from inventory. Some retailers have decided, why do I need to do that if I have a good partnership with my vendor? I will share with the vendor my model stock profile and I'll share those point-of-sale data with him. Every time he sees me sell one he can replace it. In a sense, the vendor writes the purchase order. Some retailers have taken the process a step further and have gone to vendors with the message: "I'm giving you this much square footage in my store. You manage it. I want this amount of return but you manage the space and decide how much inventory to carry."

Sharing of data through EDI opens up new worlds of possibilities. For instance, what can the apparel manufacturer do with the data? The manufacturer consolidates the data from the retail stores and analyzes them for sales trends to do forecasting and advanced planning in his own plant. Note that the advantage gained here is time. Data sharing compresses the time between point (and time) of sale and the vendor's knowledge that the sale took place.

Milliken and other firms in apparel are using the coding technology to bypass distribution centers and delete another time link from the chain. With immediate feedback on what is selling, goods can be shipped directly to the store. If it is neces-

---

[4]Remarks by a textile industry representative, "Time-Based Competition Conference," December 1988.

sary to go through a distribution center, Milliken prepacks it by store locations so that it can be cross-docked and onto the selling floor faster. The whole effort focuses on shrinking the time to the selling floor. There has to be benefit in it for everybody. It is not a camouflaged campaign to move inventory from the retailer back to the vendor; instead, it is a campaign to time-compress the entire cycle.

To see how EDI can eliminate the economic waste of a distribution center, consider a recent program by Milliken with area rugs. In this case Milliken supplied the finished product, an oriental-design area rug to a major retailer. Milliken made the following offer to the retailer: "You send us the daily orders that you get from the consumer, we will manufacture the rugs and ship UPS directly to the consumer's home." This meant that the retailer could eliminate not only his entire distribution center for that product, but all of his inventory except for display items in the showroom. The retailer responded, "That will work only on regular sales, but commercial sales are such a spike in our sales pattern that we've got to carry inventory to cover the spikes." Milliken responded by asking, "What percent are the spikes?" It turned out that two-thirds of their sales were promotional. If they guessed wrong on those two-thirds of their sales, they had excess inventory. Milliken then said, "It's all or nothing. It's either quick response or it's not quick response; we will satisfy your orders. Now you have to tell us when these big spikes are coming. We have got to share that kind of information." The retailer agreed and now Milliken is taking orders on a daily basis and shipping directly to the consumer. The system responds much faster than through the traditional DC channel and with a fraction of the inventory. Moreover, the retailer's costs were slashed by 13 percent.

In tailored clothing, an apparel segment in the United States not noted for speed, a time-based competitor has emerged that bases its comparative advantage on a novel blend of data interchange and computer technology. Custom-Cut Technologies, Inc. (CCT), a firm based in Cleveland, Ohio, has developed a patented system for custom-made, tailored suits that captures the essence of quick response: a link between faster manufacturing cycle times and innovative uses of information technology. With selected retailers, they have installed the following system: the customer is measured for the suit in the retail store, the customer's home, or his office. Measurements are transmitted electronically to a PC and transmitted by modem to CCT's manufacturing facility in Cleveland. Computer-aided design software in Cleveland converts the measurements into a cutting pattern which is fed to an automated laser cutter. The fabric is manually stitched using computer-generated instructions. The finished suit is shipped to the retailer, at which time minor alterations are made (and the customer's measurements are stored in the data base for future reference). In 1989, turnaround time was three weeks; CCT's goal is 7 to 10 days, by improvements in the assembly schedule.[5]

REDUCING LEAD TIMES FOR FASTER MANUFACTURING RESPONSE

Better, more timely information told the apparel partnerships how to respond to the customer, but traditional apparel manufacturers were hobbled by long lead times and were unable to respond quickly. To turn knowledge into action, therefore, a second major thrust of the QR campaign was aimed at reducing manufacturing cycle times. The product resupply line must shrink to make QR a reality.

In attacking lead times in the apparel industry, the QR movement encountered the same obstacles that frustrate JIT implementations in traditional manufacturing—specifically, managerial performance measures based on direct labor and machine utilization, rather than time- or quality-based measures. Faced with these incentives, manufacturing managers react predictably by scheduling long production runs in large batches. Large WIP inventories, slow reaction times, and quality/rework problems, then, are the logical results of a focus on direct labor.

The situation observed in 1987 at a Tennessee boot manufacturer is typical of the apparel industry's problems. Walking across the factory floor was difficult because aisles were blocked

---

[5]Theodore W. Schlie and Joel D. Goldhar, "Product Variety and Time Based Manufacturing and Business Management: Achieving Competitive Advantage Through CIM," *Manufacturing Review*, 2, no. 1 (March 1989), pp. 32–42.

by large queues of WIP inventory at each work station. The normal order took over three weeks to wend its way through the shop. The plant supervisor was not overly concerned with the situation; it had always looked that way to him. He commented, "My job is to keep these machines running, and so we have to make sure the operators always have enough work on hand. That's the way they get paid. That's the way we've always done it." In this plant, the major obstacle to achieving faster response was the inertia built up through years of emphasis on standard costs.

The early QR developers cleverly avoided a confrontation over the cost issue with an effective diversionary attack: They took steps to expose the high percentage of nonvalue-added time in apparel manufacturing. They adopted a JIT approach by focusing on increasing value-added time instead of adopting a frontal attack on cost and direct labor. Quick response moved to reduce lead times (and cost, indirectly) by striving to eliminate all the nonvalue-added steps.

To draw attention to the problem, KSA first carried out an operations analysis in which it isolated the value-added time in the production of different types of garments. In at least 99 percent of the cycle time for a typical garment, KSA found, on average, that no value was added. For example, in the manufacture of a T-shirt: total time in the manufacturing cycle equaled three days, of which three minutes (or 0.2 percent of the time) was devoted to value-added operations. For a more complicated product, tailored clothing, the traditional lead time was six weeks, of which KSA found that two hours (0.8 percent of working hours) was value-added time. Clearly, a sizable window of opportunity for lead-time reduction existed in apparel manufacture.

To reduce lead times in apparel, the developers of QR launched a two-pronged attack to reduce waste and increase the percentage of value-added time. The key changes were modular manufacturing, followed by smaller batch sizes. In fact, modular manufacturing helped bestow the manufacturing flexibility that made smaller batch sizes economically feasible.

Modular manufacturing requires a physical change in the layout from a traditional process, or job-shop, orientation to a

EXHIBIT 4
Modular versus Traditional Layout

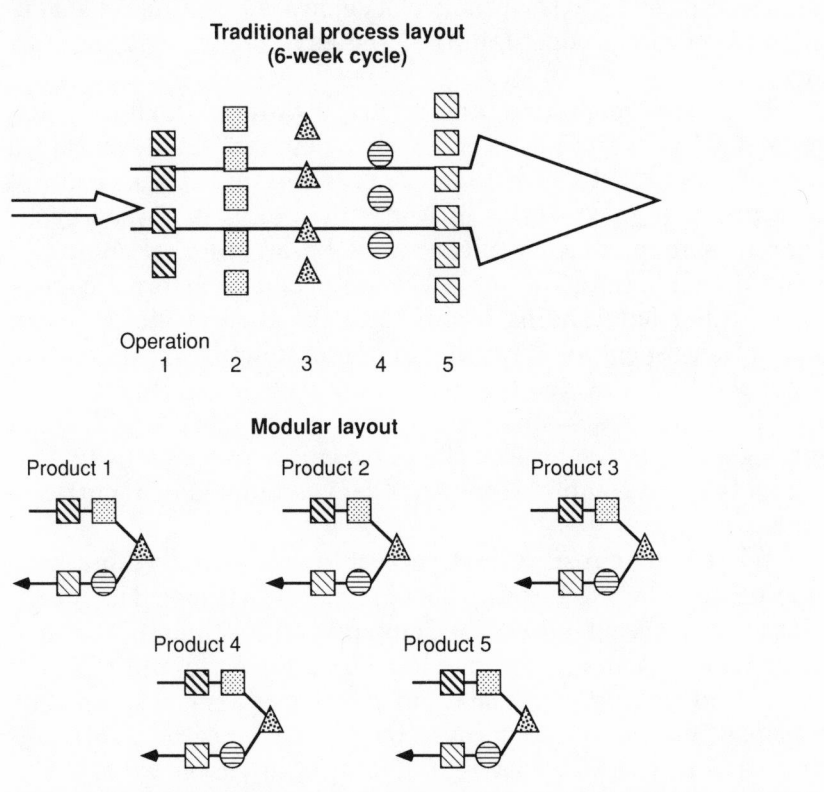

Source: Kurt Salmon Associates.

product-flow arrangement. Exhibit 4 displays the changes in layout and flow in converting a plant that makes five different garment styles from a conventional job shop to modular manufacturing. Instead of sending all styles, sizes, and fabrics of a garment through the same machine groups, the layout is altered to accommodate small, slow lines dedicated to a specific style. For a boot manufacturer, a separate flow line could be set up for the assembly of boots with exotic skins, another for boots with special heels, zippered uppers, and so forth. This layout would avoid having a jumbled flow of orders through each processing area. In other industries these layout and process flow changes are known by different names, such as "cellular manufacturing," which refers to the practice of separate machine cells dedicated to specific product families, and as a "factory within a factory."

How does modular manufacturing provide flexibility and quicker response to shifts in market demand? Suppose that a garment manufacturer with a conventional layout (like Exhibit 4) is producing five styles of knit shirts on a five-week cycle. Therefore, to make some of each style for an order (or change in an order) could take four to six weeks depending on order composition, other outstanding orders, and the current status of the production schedule. With modular manufacturing, individual U-shaped lines devoted to a particular style make it possible to produce some of each style for an order on any given day. Instead of responding in five weeks, the plant can revise its schedule on a daily basis. A complete order could be assembled in a matter of hours.

What kind of results have been obtained with modular manufacturing? Gerald Littell, a senior manufacturing vice president for The Joseph and Feiss Company, states that "We applied value-adding management/modular techniques to minimize cycle time in cutting. It reduces our piece goods risks, speeds our response time, lowers our investment, and helps motivate employees better."[6] The boot manufacturer to which we referred earlier has reduced its average production lead time by a week through the installation of modular lines for certain higher volume boot styles. More specifically, the chart in Exhibit 5 displays the results of modular manufacturing for different types of apparel, reported by Kurt Salmon Associates.

Modular manufacturing should be viewed only as a transitional stage in a progression toward more flexible manufacturing systems. Modular lines are the QR equivalent of the "focused factory" concept described in Chapter 2. They provide a short-term flexibility fix while the firm grapples with the more diffi-

---

[6]"Quick Response Implementation: Action Steps for Retailers, Manufacturers and Suppliers," Report by Kurt Salmon Associates, 1988.

EXHIBIT 5
Apparel Industry Results with Modular Manufacturing

Product	Cycle Time		Space Reduction	Quality in Percent Defective	
	Before	After		Before	After
Tailored clothing	13 days	2 days	12%	5.3%	1.0%
Knit shirts	4 days	10 minutes	25%	2%	1.2%
Dress shirts	3 days	1 hour	50%	2%	0.2%

Source: Kurt Salmon Associates.

cult problem of reducing changeover times between styles. With reduced changeover times, batch sizes can be cut and different styles can be manufactured on the same line without the diseconomies of small-lot production.

In addition to modular manufacturing, the QR movement also looked to technology to reduce manufacturing cycle times. For example, consider the fabric order-processing sequence. At Milliken, the traditional medium for transmitting an order was the U.S. postal service, perhaps the epitome of slow response. Today, order information is transmitted over computer wires, over telephones, and computer to computer. On dye and finishing fabric at Milliken, a four-week cycle has been reduced to one week in some product offerings. In apparel design, computer-aided design (CAD) systems shorten lead times for apparel designs, samples, and patterns. The CAD systems reduce lead times in two ways: First, they significantly increase the productivity of the designer; second, the design data are stored electronically and can be instantly transmitted to others.

Subsequent steps in lead-time reduction under QR should follow the JIT pattern outlined in Chapter 2: reducing production batch sizes, moving work stations closer together for both space and time compression, and increased employee involvement in work teams. As in JIT, quality usually emerges at this point as a critical problem. For instance, at the Tennessee boot manufacturer, as inventory was reduced and a discernible product flow emerged out of chaos, management was shocked to discover how many batches of boots were being recycled to earlier stations to correct quality problems. The firm had been pleased with the defect rate on product shipped to customers, but heretofore the serious quality problem within the plant had been hidden by the jumbled flow and mountains of inventory: The quality "rocks" were hidden under a "lake" of WIP inventory. Having learned that excessive rework on the lines was contributing to long average throughput times, the firm began a quality-awareness campaign. As a first step they began to collect and report performance measures on quality at different stages of the process to pinpoint the source of defects. Almost immediately, quality began to improve, the backward flow of rework diminished, and the average throughput time shrank.

SMALLER, MORE FREQUENT, ORDERS

One of the goals of QR, like its JIT counterpart, is smaller, more frequent orders—that is, a flow of smaller batches through the system. Accompanying this flow must be more frequent interchange of information between the partners; information, too, must be exchanged in smaller bundles. As a result, QR attenuates the large surges that occur with monthly orders when partners fail to share daily information about what's selling and what's not. Consider the analogy of a snake consuming an egg. A snake can crack open its jaws and swallow a whole egg. The apparel flow system, operating with monthly orders, is like the skinny snake with a gigantic egg in its throat. The egg is absorbed eventually, but the lump moves slowly and uncomfortably through the system. On the other hand, the goal of QR is to break up the egg into tiny pieces to feed the snake; the process works more smoothly and less painfully with frequent, smaller orders.

Modular manufacturing and faster data transmission make small-batch transfer of apparel feasible throughout the supply chain. With JIT linking textile to apparel to retail, retailers can carry less stock, apparel firms can produce and ship in smaller lots, and the textile manufacturer can schedule shorter, targeted fabric production runs. As Thomas O'Gorman, president of

Greenwood Mills, explains, "Our average dye lot [before QR] was 120,000 yards; today it's 11,000."[7]

As an example of QR in action, consider a recent pilot program in ladies' slacks in which Milliken participated as the textile partner: "In this program we were tracking 18,000 pairs of slacks in 225 stores. We looked at the sales after the first four weeks of sales. It is very important to look at sales early because if you wait until the end of the program, what happens? Everything you bought was sold, perhaps as a markdown, but sold nevertheless. You've got to look at the data, for both color and size, while there is a full offering in the store. In the slacks program, we saw early on that size 10 had far too much inventory and was headed for serious markdowns. On the other hand, a lot of sales were going to be lost in size 16. In the same program, a color analysis revealed problems with charcoal. Before the program started buyers wanted to drop the color because sales were forecasted very low, at only 4 to 5 percent of sales. In addition, charcoal was the most expensive color, and the retailers wanted to have fewer colors in the line. After the first four weeks, we found to our surprise that charcoal was the third highest selling color. With quick response we were able to get the goods dyed and back on the selling floor quicker in the right colors and sizes. These are not difficult analyses. But few people have the short-cycle manufacturing and information interchange to do it."[8]

THE FINANCIAL PAYOFF

Quick response pays bonuses for all three players in the game—textile, apparel, and retail. In basic merchandise, KSA reports increases in return on assets from an industry average of 9 to 12 percent in textile manufacturing, from 6 to 13 percent in apparel, and from 7 to 12 percent at the retail level. Significantly, the gains by one partner do not come at the expense of another; there is financial synergy in a QR linkage. According to Ed Haggar, Jr., president of Haggar Women's Wear, "For a retailer to

---

[7]"Quick Response Implementation," Kurt Salmon Associates, 1988.

[8]Remarks by a textile industry representative, "Time-Based Competition Conference," December 1988.

earn the same margin contribution from a non-quick response supplier, he has to negotiate a 25 to 35 percent price discount."[9]

In the ladies' slacks program mentioned earlier, Milliken found that, in 1986, the year prior to QR, 38,000 units were sold in 89 stores with a non-QR program. One year later, with the same item and a QR program, 97,000 units were sold—yielding a 256 percent increase in sales.

The effects of QR have even been felt in the notoriously slow furniture industry. In a Milliken-sponsored program with upholstery fabric, the range of colors in fabric supplied to the manufacturer was increased from 3 to 20. The goal was relatively quick response: six weeks. This does not sound like much unless you have recently ordered furniture; three to four months is the industry average. For the QR program the best-selling recliner chair was selected, the consumer was offered a wider range of colors, and the price point was moved up from $299 to $399. However, the consumer could get the chair in the desired color without having to wait months to receive it. The chair continues to be the best seller after 17 months, and profits have soared.

The greatest potential for QR lies in fashion merchandise, where everything depends on being close to the customer. With the systems in place and the bugs worked out through programs in basic merchandise, only incremental investment is required to roll out QR in fashion. In pilot studies, KSA reports that retailers have garnered 35 to 40 percent gains in fashion, primarily by minimizing stockouts and lost sales. Inventory turns increased from four times to over seven times per year at retail; inventory investment decreased at every point in the supply chain.

HEARTS AND MINDS: THE KEY TO SELLING QUICK RESPONSE

Apparel retailing has undergone massive restructuring in recent years. Mergers, acquisitions, and enormous leveraged buyouts have brought tremendous pressure for cost cutting in

---

[9]"Quick Response Implementation," Kurt Salmon Associates, 1988.

retail to generate the cash flow necessary to forestall drowning in a sea of debt. At the time of the Time-Based Competition Conference in December 1988, the view was expressed by participants that QR alleviates the cash flow problem by helping retail firms trim inventories and slash operating costs. One industry expert stated, "We can ensure that when a consumer walks into the store, she can find the style, size, and color she wants, and we can do that with less inventory. Some pilot studies that were done early on proved that it worked and put some hard quantifiable numbers on it."[10] For some large retail chains, unfortunately, the message did not get through fast enough. The enormous debt assumed to finance the Campeau takeover of Allied and Federated Department Store chains has forced them into Chapter 11. Quick response might have helped stave off bankruptcy; it can certainly help others avoid a similar fate.

All the technological obstacles to QR have been removed. In the early years there were a number of thorny technical issues, but with the MIS gurus taking the lead, industry standards for barcoding, scanning, and communication via EDI are being adopted rapidly. Several years' experience with successful partnerships has shown that it works. Nevertheless, much of the apparel industry watches and waits. The challenge today is to take QR from an MIS-driven function to a merchant-driven responsibility.

Hearts and minds must be changed if QR is to permeate the apparel industry. Buyers and their managers have typically been measured on gross markup and not on total cost. Consequently, their vision rarely extends beyond their own departments. A radical shift to an emphasis on total system cost and system response time takes nothing less than a cultural change within an industry. Retailers must look to suppliers not as contractors, but as partners in a time-compression and cost-reduction process.

Cultural change must come from the top (see Exhibit 6). Incentives and performance measures within the firm must be changed to support QR, and those decisions can be implemented only at the highest level. A synchronized supply chain partner-

---

[10]Remarks by a textile industry representative, "Time-Based Competition Conference," December 1988.

EXHIBIT 6
Keys to Quicker Response

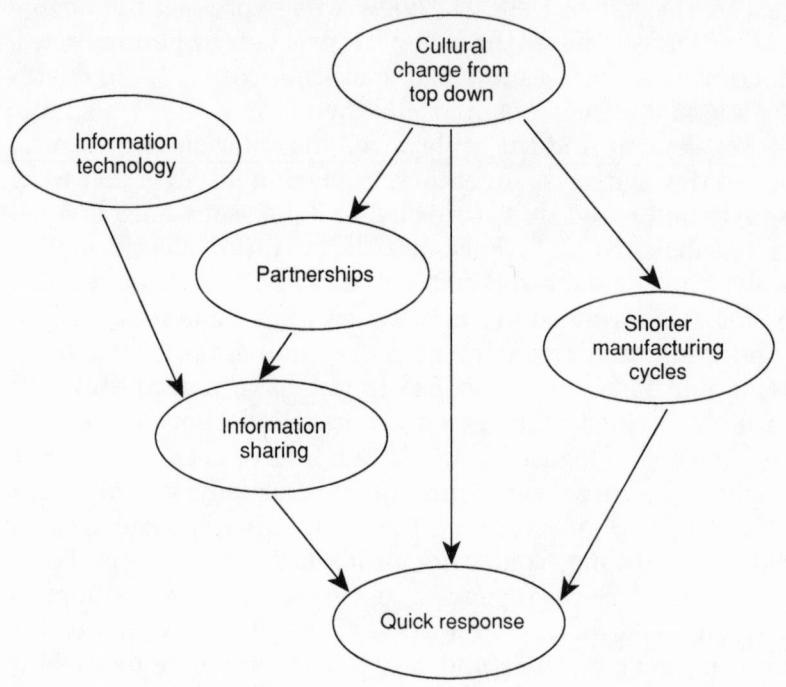

ship can be established only by agreements among the top executives of each firm. Not surprisingly, the pioneering firms in the
QR movement, such as Milliken and Wal-Mart, have all had
strong leaders who were firmly committed to the concept.

The QR movement provides an excellent model for would-be
time-based competitors in all industries, not just apparel. Essentially, QR is the extension of JIT principles to an entire value-
delivery chain. In manufacturing, firms that have undertaken
JIT have typically focused, myopically, on their own processes
or the direct link with a vendor. Few industry groups have had
the vision or the cooperation to synchronize the total supply
chain from raw material to customer. The evidence from the QR
adopters conclusively demonstrates that it offers a tremendous
competitive advantage for those who embrace it. Roger Milli-

ken, chairman of Milliken and Company, states a view that applies to all U.S. manufacturing: "Companies that learn how to make quick response work will be the winners in the 1990s."[11] Those who do not embrace it are the odds-on losers.

---

[11]"Quick Response Implementation," Kurt Salmon Associates, 1988.

PART 4

THE TOP MANAGER'S ROLE

CHAPTER 12

THE CEO'S ROLE IN TIME-BASED COMPETITION

Dan Ciampa

Editor's Note: Dan Ciampa is president and chief executive officer of Rath & Strong, Inc. in Lexington, Massachusetts. He is the author of *Manufacturing's New Mandate: The Tools for Leadership* (Wiley, 1988). Since joining Rath & Strong in 1972, Mr. Ciampa has specialized in the people side of problem solving; he has helped leaders better manage organizational cultures and implement new business strategies. In this chapter he discusses the role of top management in creating the organizational environment needed to be a time-based competitor.

Achieving time-based competition demands a commitment from people throughout the organization to succeed. Application of the techniques that can reduce lead time and make the organization more flexible and responsive requires all personnel to go out of their way to solve problems, to do the right things in the right way, and to operate in a team-oriented way. Employees must want to work in these ways; their commitment is foremost in a company's ability to compete on the basis of time.

A firm's effort to compete on time raises four different issues with which top management must deal. The first issue is that of integrated manufacturing, or how to integrate and put into the right sequence the various tools and techniques to take time out of the production process. These tools and techniques come from Total Quality Control (TQC), Just-in-Time (JIT), and computer-integrated manufacturing (CIM). Second, important people-related issues arise while implementing total quality and, in particular, JIT. Third, certain factors of organizational climate and company leadership tend to determine whether those people-related issues bring opportunities or bring problems. Last, a well-managed implementation sequence is needed to create an environment in which time-based competition can become a reality. By dealing with these issues properly, a firm can foster the development of the all-important employee commitment to time-focused processes.

INTEGRATION OF TQC, JIT, AND CIM

Modern integrated manufacturing is based on the total quality concept—namely, giving customers what they want without making mistakes in the process. Total Quality Control provides a firm with a climate in which innovation thrives and root-cause problem solving prevails. The elements of this climate are explored later in this chapter. TQC also provides a predictable process. The likelihood of achieving the enormous opportunities available through TQC and JIT efforts is much greater when the process by which the product is produced and delivered to the customer is predictable—not necessarily perfect, but at least predictable. Predictability of this kind is a prerequisite to the flexibility advocated by George Stalk of the Boston Consulting Group in Chapter 3. Finally, TQC provides an intense customer orientation by emphasizing "doing it right the first time" so that efforts are not repeated. TQC reverses the dictum that a firm never has time to do it right the first time, but always has time to do it again.

The customer element of TQC is comprised of two parts. One is the "external" customer. In total quality efforts that have been around for several years, employees who previously did not interact with customers now understand customer needs to a greater extent. Surveying customer needs, bringing hourly workers and engineers together with key customers, and making sure top management interacts with counterparts in customer companies all improve the external customer relationship. The second part is the "internal" customer. This means that employees care about the quality and output of each other's work and agree to give what is needed, when it is needed, and in the form that is most helpful to performing the job. Employees treat each other as the whole company should treat its most important customer. Making quality important to every employee is probably the most evident part of the total quality effort. Histograms, scatter diagrams, and other methods of displaying data on performance help in this regard. Companies that respond well to customer demands for better quality and delivery periodically post such charts so employees can know how they are performing relative to customer needs. Other companies bring employees together periodically to trumpet and celebrate successes in performance.

Total quality control also pushes decisions downward and that requires firms to increase the involvement of employees at the middle and lower levels in the decision-making process. One of the most important forms of this involvement is problem solving. Dr. Joseph Juran, a guru in quality control, has said for over 50 years that the way to improve quality throughout an organization is on a project-by-project basis. This approach should be supported by analytical techniques like pareto analysis and cost-of-poor-quality analysis as well as by education and a lot of training of task teams. Groups of employees can work together to analyze and solve problems they encounter day in and day out without depending on specialists from corporate departments. Employees should have the responsibility for solving the problems that affect their work. For this approach to work, however, those employees must be trained in the right techniques and be given the power to implement changes.

Total Quality Control provides the base for company-wide, continuous improvement in today's manufacturing firms. Once that base is established, JIT systems can bring enormous gains. Although JIT alone brings gains, when JIT is implemented alongside TQC, the gains are staggering: 80 and 90 percent reduction in lead time, 75 percent reduction in setup time, 60, 70, and 80 percent reduction in inventory, and 60 percent reduction in indirect labor and space savings. In some companies, cost of purchased parts has been reduced by 50 percent. Not enough research is available, however, to compare the results of JIT and TQC together with the results of JIT alone.

While JIT means doing it right the first time, it further means doing it the first time *only* if it adds value. Continuous elimination of waste—for example, in counting the product, storing it, and carrying it around—is the key objective. Use of JIT techniques requires a firm to produce one item at a time and only when it is needed—and to conserve time, push decisions to lower levels, and simplify the process.

JIT systems concentrate only on activities that add value; wasted steps are removed from the process, a necessary action to improve flexibility. Two Rath & Strong rules of JIT implementation speak to this point. One is the *3 to 20 percent* rule; the other is the *1/2 of 1 percent* rule. The firm has analyzed many different work processes by documenting the steps for the product or piece of information—how far it travels, how long it waits, and so on. The results always showed that only 3 to 20 percent of those steps added value to the product or piece of information, and, moreover, those value-added steps occurred less than 0.5 percent of the total cycle time. Time and again—whether it was an insurance company, bank, engine manufacturer, or soft drink manufacturer—these rules were borne out.

What about the 80 to 97 percent of the steps that do not add value to a product? They generally fell into three categories. One kind of nonvalue-added step was needed because of a mistake that required time to undo or redo. A second kind of nonvalue-added step happened because no one thought through a process differently. A 75 percent reduction in set-up time without capital expenditures on equipment is a realistic goal only when the firm thinks through the process in a different way. A third kind of nonvalue-added step happened because the company lacked a better way to perform a set of tasks. Computer-integrated manufacturing and systems integration can aid in solving this problem.

Xerox is an example of how attention to nonvalue-added tasks can bring dramatic improvement. Xerox eliminated incoming inspection on 90 percent of its product. While the company eliminated inspection paperwork, it also trained employees from indirect labor to become vendor auditors, while simultaneously reducing its vendor base. Xerox reduced product cost by 45 to 50 percent. Air Products, Corning, Hewlett-Packard, Pepsi-Cola, and John Deere are other companies that eliminated substantial time and cost from their processes by concentrating on removing waste from nonvalue-added steps.

Once waste and redundant systems are eliminated by JIT, the firm's next challenge is to automate and integrate the revamped systems. In this phase, the application of computer-integrated manufacturing (CIM) can add efficiency and predictability. Joe Harrington of Arthur D. Little first coined the term *computer-integrated manufacturing*. Basically, Harrington said, "Let's look at manufacturing from the time the product is conceived until the time it gets into the customer's hands, and determine all the information needs that are associated with that flow." The point of modern integrated manufacturing is that companies can implement three different sets of tools and techniques—TQC, JIT, and CIM. Each set contributes something different. A certain interdependence among these tools, however, implies an ideal sequence by which they can be implemented. First, TQC is used to lay the groundwork inside and establish a customer orientation, then, JIT to ensure flexibility and eliminate waste and time, and, finally, CIM to bring added efficiency and predictability through automation and systems integration. Frequently, these three sets of tools and techniques compete with each other for management attention. That competition diminishes the power that can result from their combination. Ways to implement such combinations are discussed in a later section.

PEOPLE-RELATED ISSUES IN TQC AND JIT

Companies implementing TQC and JIT frequently confront significant people-related issues. These issues represent dramatic changes in the way the business operates. Each change poses certain opportunities and dangers for the firm, as summarized in Exhibit 1. First of all, the employee's job content will become broader and require new areas of competence. The benefits are

EXHIBIT 1
Five Major Changes on the People Side

Change	Opportunities	Dangers
1. Job content change	■ More growth opportunities	■ More unemployment ■ Worker discontent
2. Better interpersonal communication	■ More face-to-face contact ■ Better working relationships	■ Worse working relationships
3. Roles and relationships change	■ Enrich jobs and offer more challenge and satisfaction ■ Better control decisions	■ More decisions not made ■ More confusion ■ People not able to adapt
4. Decision-making authority pushed down	■ Greater motivation ■ Faster decisions	■ Threatening ■ Vertical checks and balances
5. More user involvement	■ Better meet needs ■ More commitment	■ Longer development time ■ Complicate the process

Source: Rath & Strong, Inc.

that more growth opportunities are created, and jobs can be en-riched. Dangers, however, also come with such change. For ex-ample, employees can be displaced or become unemployed or merely unable to perform if the right training programs are not created.

A second people-related issue is expanded interpersonal communication, especially across departments. Most companies have grown in a compartmentalized way, where, for example, employees in engineering, especially at the middle-management level, rarely communicate with those in manufacturing or mar-keting. Currently, simultaneous engineering efforts are improv-ing this, but in many cases, the improvement is coming too late. Benefits that come with cross-departmental communication in-clude face-to- face contact and better working relationships. But the danger exists that working relationships may falter precisely because of more face-to-face contact. The outcome depends on how companies prepare and train employees to work together in new ways.

A third issue is that employees are required to behave differently because their roles and responsibilities are changed. This change brings an opportunity to enrich jobs and offer more challenge and satisfaction. But, the dangers inherent include these: more decisions falling through the cracks, more confusion as to who is to do what, and employees unable to adapt well to new roles.

These changes in employee roles and relationships require employees to act differently in at least four ways: (1) employees must collaborate more frequently with co-workers because of the relatively large number of people involved in decision making, and this, in turn, requires higher trust; (2) employees must initiate action rather than rely on being told what to do because one important element of time-based competition is fast decision making—the strategy simply allows no time to wait for directives to flow through traditional bureaucratic channels; (3) because of the need to have employees take responsibility, they must develop self-confidence and become willing to take risks; and (4) employees must assimilate different pieces of data and merge them into today's mode of operation. The pace of development will accelerate and create even more options for doing work faster, and workers will be required to merge these new developments into their day-to-day activities. All of these new behaviors are required in a TQC/JIT environment.

A fourth major people issue is that decision-making authority is pushed further downward. As processes, systems, and new techniques become available on the shop floor, the employees using those techniques are asked to make decisions about how to make them work better, thus offering opportunities for greater motivation and higher levels of satisfaction. Faster decisions also can result. The danger here is that middle-management employees could be threatened as top-management employees push responsibility down to the shop floor and office workers. Vertical checks and balances that slow down the decision-making process may result. When the decision-making and time-reduction processes are out of step, time-based competition will not last. Shortening time on the shop floor by eliminating all nonvalue-added steps leads to quick gains. But ensuring that those gains last over time is the hard part. The key is to have decisions made at the point in the process (and at the level in the organization) where the problems and situations occur in the first place. The workability of such decision making depends on whether employees work together, whether they trust each other, and whether the time taken to make those decisions is reduced along with the time required to make the product.

A fifth issue is increased involvement of those employees who will use the new equipment and systems. Although greater involvement can increase the commitment of employees expected to use these methods and equipment, it can also increase development time by complicating the process of designing and engineering new products.

IMPACT OF ORGANIZATION CLIMATE AND LEADERSHIP

Companies can control whether opportunities or dangers result from these five people-related issues. Two bands run through each issue: one is the organization climate, and the other is the leadership style of top management. How well an organization—in particular, its leader—manages these two bands determines whether these issues will create opportunities or dangers.

Elements of Climate

Experience and empirical research at Rath & Strong point to six elements of organization climate that correlate most closely with TQC and JIT success (shown in Exhibit 2). When these six elements exist and an organization takes advantage of them, time is reduced, quality is improved, and customers are more satisfied. One element is influence, or the degree to which employees believe they have control over changing the context around them. Resistance to change is higher in organizations where employees have a low sense of influence because they have little ownership in the change process. That is why influence is the most important of the six elements.

EXHIBIT 2
Climate to Support Total Quality Control and Just-in-Time

- Influence
- Responsibility
- Innovativeness
- Desire to change
- Teamwork
- Common vision

Source: Rath & Strong, Inc.

A second element is responsibility—the degree to which employees take responsibility for changing what is around them without being told what to do or how to do it. The higher the self-perception of influence, the more likely employees are to assume this kind of responsibility. Companies with low levels of employee responsibility tend to need JIT more than companies where such responsibility runs high. Implementing JIT is more difficult in those companies, however, because employees are required, as part of their work routine, to use initiative and go out of their way to solve problems that occur.

A third element is innovation, which is the formulation of new ideas. High innovation exists in companies in four different forms. First, some companies encourage trial and error. In companies utilizing high levels of experimentation, the sense of innovativeness will also be high. Second, companies accustomed to working in small teams will create a cross-functional task force to solve a problem. Third, some companies encourage innovation through "champions" of new ideas. In some firms, finally, the leader pushes new ideas. For instance, consider how Chrysler introduced a convertible into its new line. Lee Iacocca thought that Chrysler needed a convertible. The chief engineer said, "It will take nine months to do that." Iacocca responded, "No, you don't understand. Take a car and saw the top off the damn thing and do it now." Iacocca then drove around Detroit in this convertible and counted the number of waves he got from onlookers. When he had seen enough, he drove back to the plant and said, "Build it." This case represents purposeful intention from the leader, which may be the most important factor in innovation.

A fourth element of organization climate is the desire to

change the status quo. If employees are satisfied with conditions around them, they will not accept the pain associated with the changes that are necessary to reduce large amounts of time and waste. Tapping into employees' desire to change is perhaps the most important job of the leader when moving into time-based competition. In doing so, the leader must understand the "phenomenon of constructive change." (This phenomenon is represented graphically in Exhibit 3.) What tends to happen is that dissatisfaction inside the organization raises the likelihood that constructive change will occur. Dissatisfaction is a call from employees that they are ready to make changes in how things are done. If top management does nothing to respond to that readiness, the employees' emotion turns into frustration. It is important for the leader to understand this phenomenon because an organization should do different things to respond to dissatisfaction than to frustration. If an organization needs to deal with dissatisfaction, management should seek to motivate achievement. It should set goals that are realistic but challenging, give constant feedback on the achievement of those goals, and make

EXHIBIT 3
Constructive Change Phenomenon

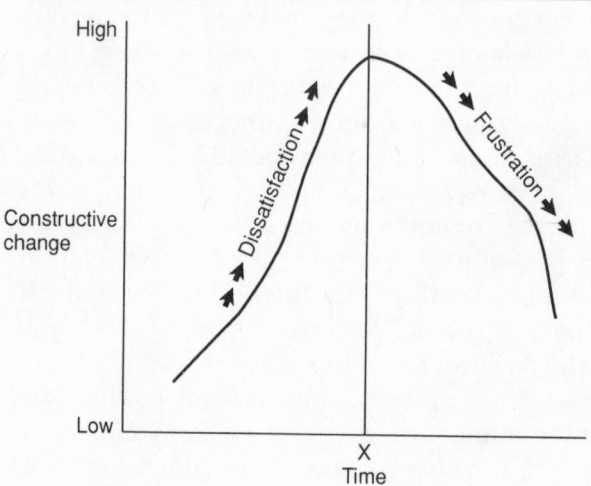

Source: Rath & Strong, Inc.

the reward system consistent with the behaviors expected. Management should hold employees accountable for those goals and make sure the right information systems are in place. These tactics, however, will not alleviate frustration within an organization. Instead, employees become more frustrated. Frustrated employees need a vision of the way things could be. Achievement-oriented actions like goal setting and feedback allay frustrated employees *only* when the employees possess a vision of how things could be.

A fifth element of climate is teamwork. Teamwork means that personnel work together in harmony and subordinate personal gain to achieve common objectives. For a TQC/JIT environment to thrive, teamwork within departments is important, but cross-departmental teamwork is even more vital.

The last climate element is common vision. What that means is that employees throughout the organization have a mental picture of how things could be. *Vision* has become a generic word in today's organization lexicon; this is not how we use it here. In the present context, vision means a mental picture of how the company will look and how it will feel to work there in the future—that is, what a business that competes successfully on the basis of time will be like. A vision is not a mission statement and not a set of goals, but too often today that is what results from vision-setting activities.

The leader's role in creating a common vision is most important. Whether the vision becomes commonly held or not depends heavily on the way in which the firm's leader goes about influencing employees. The leader can use at least three influence styles: assertive persuasion, rewards and punishment, and common vision. Assertive persuasion is a style that depends on facts and logic ($A + B = C$ and, therefore, the only option is D). The rewards-and-punishment style says, Do what I want you to do; if you do, I'll give you something you want; if you do not, I'll take away something, or not give you something you want. The first two styles are not uncommon among American leaders, but neither one is particularly effective in establishing a widespread, commonly held vision of the way things can be that excites and motivates employees. The third option, common-vision leadership, is more effective.

Common-vision leadership is the ability that some leaders have to paint a compelling and exciting picture of the future. That picture is usually based on values that employees feel are important. The essence of a common-vision-based style is enabling employees to have a mental picture of how things can be that enables them to make that picture their own. It is not important that every employee describes the picture in the same way. In fact, the visioning exercises used by some companies err by over-codifying. Employees will commit themselves to a vision only if they can make it their own by putting their own particular spin on it. The art of common-vision leadership is to foster personal interpretation of the vision, but still ensure that various interpretations are not inconsistent, so that all are moving in the same direction.

Influence, responsibility, innovation, desire to change, teamwork, and common vision—these are the six elements of the organization climate. When they exist, the likelihood is significantly greater that TQC and JIT will succeed in creating an organization that is fast and flexible.

Top Management Leadership

The other band that determines how companies handle people-related issues is organizational leadership. TQC, JIT, and CIM can be combined to create flexible, time-based, competitive organizations. But these innovations, by nature, are cultural changes. Their ability to produce a competitive edge for a company depends on whether they result in broad, permanent changes to the organization's culture. Corporate leaders cannot delegate the management of the organization's culture. How leaders manage an organization's culture usually determines the success or failure of any change effort. Management leaders, therefore, must understand their styles, their strengths and weaknesses, and what elements will complement their styles to support their strengths and compensate for their weaknesses.

There are many ways to codify leadership style. One is offered by Alex George, a foreign policy advisor to U.S. presidents dating back to Franklin Roosevelt. George's research says that modern U.S. presidents have exhibited three styles: *competitive, formalistic, and collegial.* Quite a bit about the task of leading an organization's cultural change can be learned by studying these particular leadership styles.

The *competitive leader* creates competition and, sometimes, conflict by knowingly assigning overlapping areas of responsibility within top management. Through that competition and conflict among managers, information is generated that enables the leader to make decisions. This style generally develops creative options, but turnover tends to be high because of the conflict it spawns. When a group of managers has to work together as a team, the style tends not to work well. Furthermore, because the competitive leader rarely delegates true authority, he or she usually becomes involved in each decision-making step. As a result, it is not uncommon for a competitive leader to have time-management problems.

A *formalistic* leader emphasizes order. Formal leaders need around them a staff superstructure that provides options from which they can choose the best course of action. They must have loyal, analytical staff aides who bring forward these options. The formalistic style tends not to work well in times of crisis because events are moving too fast and because the decisions produced often are neither highly creative nor ones that evoke strong passion from employees.

The *collegial* leader stresses teamwork and consensus building. He or she places great emphasis on collaborative work by the management team and on openness and the sharing of information. Collegial leaders are good at building consensus, creating teams, and obtaining adoption of a common vision for the organization. What these leaders do not do well is encourage dissent, and that becomes their Achilles heel. The chief danger inherent in the collegial style is the "loyalty trap," that is, employees become intensely loyal to the leader and what he or she represents. It is not uncommon for the degree of disagreement and dissent to diminish as that loyalty becomes stronger. As that happens the collegial leader enters a dangerous phase where only weakened checks and balances exist to compensate for the leader's ability to engender respect and followership. One of the key barometers of a collegial leader's success over time is the degree of dissent that is expressed. If little or no dissent is voiced in staff meetings, in formal surveys, in employee discussions, and so forth, then the collegial leader is at real risk.

It is rare for a corporate leader to exhibit only one of these leadership styles; most leaders are combinations of all three styles. Most of the leaders with whom the writer works use a mix of competitive and formal styles. While TQC and JIT require teamwork and collaboration, it does not follow that the leader must have a collaborative style to successfully implement those two systems and reap the benefits of time-based competition. While a leader may change and learn new ways of operating, a dramatic change in style—such as moving from a competitive to a collegial sort of style—is unlikely.

Realistically, leaders should recognize the downsides of their intrinsic styles, and then surround themselves with superstructures that compensate for those weaknesses. For example, formal leaders must ensure that a system exists to provide them with options for decision making. They need strong staff aides to manage information flow, team-oriented staffs, and a staff person who can build teams to mitigate the downside of a formal leadership style.

A competitive leader must balance the leadership styles on the top-management team with leaders who are formal and collegial. In this case, the competitive leader must expect conflict when collegial, formal, and competitive managers work together, but conflict is the best means for them to manage because it keeps them equal. There also needs to be a safeguard against time-management problems. Usually, that means expanding the role of an executive secretary to keep the leader's calendar; the secretary must have high tolerance for the confusion and ambiguity created by a competitive leader. The collegial leader must take steps to ensure enough dissent to safeguard against the "loyalty trap."

IMPLEMENTATION STRATEGY

A specific strategy is needed to implement a company-wide effort that combines the right elements of TQC and JIT. The objective of this strategy is to create a continuous improvement ethic throughout the organization. The strategy is based on three principles. First, any approach used to implement TQC/JIT tools and techniques that will enhance time-based competition must be tailored to the company's needs. No single method of implementation is best. Certain principles can be followed and lessons can be learned from others, but the implementation approach must be company-tailored. Second, the implementation approach must be structured, but not rigid. TQC and JIT efforts that excel have common characteristics, including the high degree to which those efforts are thought through and structured and the presence of a leader who maintains an emphasis on structure throughout the implementation process. The third strategic principle is shared responsibility. Only the leader can implement some phases, while others must be assigned to other employees in the organization.

Over the past several years, companies that have successfully implemented TQC and JIT have followed a process that is similar to a four-track model. (Shown in Exhibit 4.) The first track is the vision track. Included in the initial phase of the vision track are awareness and orientation sessions for top management as well as a process by which the top team develops a clear vision of the future. Companies that realized mediocre results in implementing TQC and JIT often failed to define a future vision. The vision clarification process should result in a mental picture of how the plant will look after a couple of years of implementation and what it will feel like to work there. The vision track, furthermore, should be fueled by an intense dissatisfaction with the status quo. Vision clarification bogs down when some members of the top team are satisfied with the status quo, have a stake in it, and seek to preserve it. This situation can only be addressed and remedied by the leader.

As commitment to TQC and JIT builds among the top team, most companies move into a strategy phase. Here, opportunities are identified: specific numbers for lead-time reduction, inventory reduction, productivity, space reduction, and other tangible objectives. The six elements of organization climate cited earlier are also measured to determine their current status in terms of influence, responsibility, innovativeness, teamwork, and dissatisfaction. A step that often occurs in the strategy phase is

EXHIBIT 4
Continuous Improvement Implementation Approach

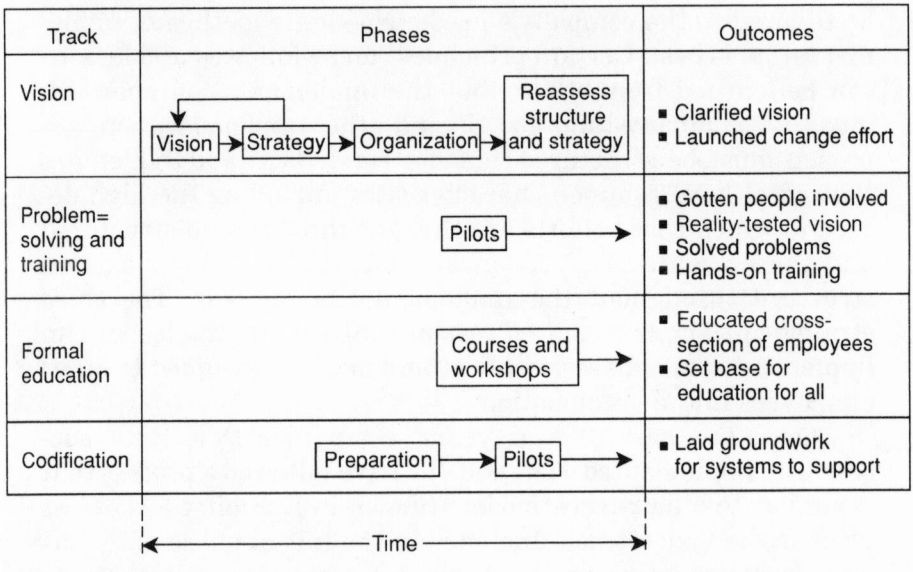

Track	Phases	Outcomes
Vision	Vision → Strategy → Organization → Reassess structure and strategy	▪ Clarified vision ▪ Launched change effort
Problem=solving and training	Pilots →	▪ Gotten people involved ▪ Reality-tested vision ▪ Solved problems ▪ Hands-on training
Formal education	Courses and workshops →	▪ Educated cross-section of employees ▪ Set base for education for all
Codification	Preparation → Pilots →	▪ Laid groundwork for systems to support

|← —————— Time —————— →|

Source: Rath & Strong, Inc.

benchmarking. The term is not used here to imply competitive benchmarking, or to compare a particular company to its major competitors. Rather, a particular firm studies companies that excel in the areas where it must improve to find what makes those businesses so good in those areas. For example, in implementing its TQC/JIT effort, Xerox asked, What are the areas we most need to improve? What companies are best in each area? Because one of Xerox's concerns was distribution, it sent a team to L. L. Bean in Freeport, Maine, to learn how that company excelled at distribution. The end result of all the work in this phase is a strategy for proceeding with subsequent phases and tracks.

In one of these first two phases, the voice of the customer must be heard loud and clear. Time can be used as a competitive advantage because, after all, the customer becomes better satisfied through quick response. The reason for higher quality and more flexibility is to give customers what they want when they want it. How satisfied customers are and what they want can be discovered by customer surveys, either formal or informal. The best time to conduct such surveys is at the start of a TQC/JIT effort and, then, at various intervals thereafter.

The third phase of the vision track is organization. During this phase, the leader of the top team must decide the best way to organize a TQC/JIT effort that will achieve the time-based competitive advantages the company seeks. Many efforts begin with a council composed of top-, middle-, and, sometimes, lower-level employees to steer the TQC effort. Sometimes that works and sometimes it does not. Some organizations achieve results with only a top team. Other organizations appoint one person as the full-time head of an effort—for example, the vice president of quality, or the JIT champion. In other companies it is best to give the responsibility to a particular line officer. After clarifying the common vision and developing a strategy of how to proceed, the top team must choose its best options for organizing its effort. Milliken, for example, selected a high-level employee, a general manager of one of its businesses, when it undertook its JIT effort. Corning, on the other hand, moved employees into full-time positions and made them heads of various development efforts. Both approaches were successful because they responded to the particular needs of the company.

The second track of the four-track model is problem solving and training. This track enables an organization to capture employee commitment by involving them in pilot projects directed at the opportunities identified in the strategy phase. A second reason for starting pilots at this stage is to solve real problems that are blocking employees from taking time out of the system. Opportunities should be piloted that are important and that can be realized. In other words, pilots need to be both visible and realistic. Some companies make a mistake by taking their most difficult problem—one that employees have struggled with for years—and using it as a pilot. This is not the best way to begin. A problem is needed that can be solved in a relatively short time. For this reason, set-up reduction is often an ideal first pilot. Set-up time on any piece of equipment can be reduced by at least 75 percent in a matter of months without capital expenditure. It is dramatic, it happens quickly, and it is visible—and that is the point of pilots. They should enable employees to see dramatic changes result quickly from using new techniques. Employees working on these pilots will talk about them with their friends and that will engender enthusiasm among the other employees.

The third track is formal education—in the form of courses and workshops—that should take place only after pilots are underway. Many companies begin their efforts with formal education, but employees can be educated too soon. While children learn best by hearing concepts and then putting them to use, adults, conversely, learn best by experiencing a process, sitting back and reflecting on it, and then developing a concept. If employees are to implement TQC and JIT, companies should first involve them in pilots. Later, employees can attend courses and develop an underlying concept of what they did during those pilots. This approach will enable employees to get more from education and to apply concepts quickly to other problems. Also, education courses must be tailored to the specific needs of employee groups rather than being generic. Having some experience before a course (whether observed or direct) and going through a session that includes cases or exercises from one's own company will go a long way toward ensuring the impact of education.

The last track of the four-track model is codification. Activities during the strategy phase often cause members of the top-management team to ask questions about the central nervous system of the business and its three constituent parts: the measurement system, the reward system, and the information system. For example, once the firm's vision is clarified, someone on the top team will ask, "Are we measuring the right things? Are we rewarding employees for the right behavior?" The point is, once pilots are underway, changing the reward system takes time. Motorola took five years before it implemented its first pilot on changing the measurement system. Milliken took seven years, Corning six, and John Deere eight. Hewlett-Packard, however, provides an example of changing the measurement and reward systems early in the process. Gary Flack was the manager of the plant that introduced TQC and JIT into Hewlett-Packard in the late 1970s. As an early step, Flack, a former division controller, developed a TQC/JIT measurement system that was approved by Hewlett-Packard's auditor, Price Waterhouse. Usually, however, the top team waits until employees have been involved in the effort for some time. Only then are changes made in the reward and measurement systems.

Another way to institutionalize TQC/JIT is to train many employees to carry on new practices that are often initiated by outside consultants or corporate staff people. Real success in reducing time and improving quality will not come unless people at a local level are able to implement the steps, changes, and innovations necessary to make it a reality.

In sum, the company leader should instigate the implementation sequence and lead the initial vision and strategy phases. The leader should then step back and allow other employees, particularly in middle management, to assume responsibility for the problem solving, pilot, and education tracks. As pilot projects are realized, the leader should focus attention on changing the measurement, reward, and information systems. Throughout the process, the leader must display his intense interest by constantly raising the right questions about implementation and its results.

Finally, whether or not a company follows the specific phases just outlined, its first mission must be to establish a vision and a strategy early in the process. Second, pilot projects must be undertaken prior to conducting educational programs. Third, changes must be made in the central nervous system of the business; changes in the reward system will require changes in both the information and measurement systems. These steps, while not the only ones that will result in positive change, are crucial to institutionalizing changes that take place.

SOME FINAL THOUGHTS

First, leaders must take their own medicine. While employees from throughout the organization must participate fully, the leader cannot expect that to happen unless employees know that the leader is acting the way he or she is urging them to act. They need to see their leader use the new tools and techniques in his or her management activities. When a decline in market share occurs or a customer-satisfaction problem arises, for example, leaders must roll up their sleeves and chart it. Employees throughout the organization should see histograms and flowcharts used by the leader if they are expected to do the same. Leaders need to take time out of the management processes so that decision making is easier and faster if they expect supervisors to do the same. Furthermore, the leader must be consistent. True commitment happens when "the going gets tough," not when it is easy. So, whether times are good or bad, the leader must be consistent in implementing the elements of TQC, JIT, and CIM.

In addition, all of upper management must act as models for other employees. The employees' enthusiasm wanes, or their dissatisfaction turns into frustration when they hear the leader say one thing and see subordinate managers do another. Likewise, mid-level managers must also act consistently. Most employees interact with middle managers and supervisors far more than they interact with the leader. When workers know the leader's vision of the future but interact daily with supervisors who act in ways contrary to that vision, their enthusiasm predictably wanes, and commitment is lost.

Last, the opportunities offered by TQC/JIT/CIM can only be achieved when employees go out of their way to solve problems that surround them by working effectively as a team. How the leader acts, what the organization climate is, and which tools and techniques are combined all affect that outcome. At bottom, the leader's challenge is to embrace the nebulous thing called organizational culture. A leader must learn to manage the motivation of employees in every corner of the company so that they are able to use the tools and techniques available to extract large amounts of time from the marketing, development, production, and distribution processes in a way that ensures the highest product and service quality. This is today's new mandate to the leader.

BIBLIOGRAPHY

1. D. McClelland, *The Achieving Society* (New York: Van Nostrand Reinhold, 1961). Lays out the basis for managing motivation.
2. D. Ciampa, *Manufacturing's New Mandate: The Tools for Leadership* (New York: John Wiley, 1988). Goes into greater detail on the topics presented in this chapter.

3. E. Hay, *The Just-In-Time Breakthrough* (New York: John Wiley, 1988). A detailed, comprehensive treatment of the tools and techniques of JIT.

4. J. Guaspari, *The Customer Connection* (New York: AMACOM, 1988). An entertaining and to-the-point description of the reason to compete on the basis of time—the customer.

5. J. Harrington, *Understanding the Manufacturing Process* (New York: Marcel Dekker, 1984). The definitive treatment of CIM from the person who first used the term.

6. J. Juran, *Juran on Leadership for Quality* (New York: Free Press, 1989). One of many books by someone who has led the quest for quality for five decades.

7. A. George, *Presidential Decision Making in Foreign Policy* (Boulder, CO: Westview Press, 1980). Research on leadership styles of U.S. presidents and their results.

CHAPTER 13

THE FUTURE EVOLUTION OF COMPETITION

Joseph D. Blackburn

In the preceding chapters, managers from leading U.S. firms explained how to develop the skill of time compression and turn response time into a powerful strategic weapon. Although time-based competition was first demonstrated on a global scale by Japanese firms, these American firms have shown that the process can be emulated. Moreover, they have shown how time can be excised from all segments of the value-delivery chain, in new-product development and distribution, as well as manufacturing.

Response time is, of course, just one factor in the competitive equation. Quality is the other important factor that has jumped to the forefront in the 1980s as manufacturing and service firms struggled to respond more quickly to customer demands. Strategic planners are faced with more than a dilemma because they must consider the firm's competitive stance along multiple dimensions—time, quality, price, and product-line variety. There are few industries today in which a firm can compete effectively by focusing on a single dimension (e.g., being the low-cost producer).

Will a new dimension of competition emerge as business and consumers move into the 21st century? Conventional views of strategy have been shaken by what appears to be the rapid emergence of time and quality as dominant factors. However, though the pace of change in manufacturing strategy seems to be accelerating, a closer look at the trends indicates that these changes have been underway for several decades. This chapter examines how competition has evolved along different dimensions and, on that basis, makes some educated extrapolations into the coming decade and beyond.

COMPETITIVE TRENDS

During my talks to corporate executives about time-based competition, two distinct groups invariably form within the audience: a group of "believers" who are convinced that time is emerging as the dominant dimension of competition and a smaller group of "skeptics" who view competing on time as another passing fad. Both groups of executives, however, are concerned about the future form of competition in their own industries. They ask, "Where will this all lead? How will competition evolve in the 1990s and beyond?" Readers of this book, at this point, presumably divide into similar groups and ponder similar questions about the role of time and the evolution of corporate competition.

Fortunately, it is possible to construct a looking glass into the future for some answers. The trends are clear and unambiguous. In virtually every industry, one can predict with confidence how competition will evolve. The best way to visualize this is to plot the norms in an industry with respect to response time, quality, and product variety. More specifically, plot a graph like the one shown in Exhibit 1. Put the 1950s, 60s, 70s, and 80s on the x axis and plot lines that roughly indicate how industry norms have changed from decade to decade during that period. In most industries—industries as diverse as eyeglasses, package delivery, wholesale distributions, consumer electronics, motorcycles, and automobiles—the patterns are identical: The graph tends to show quickening market response time, improving standards of quality, and increasing product variety. (The plots for service industries are strikingly similar.) One vital measure of response time is the time taken to develop new products. As indicated in Chapter 5, the general industry trend is

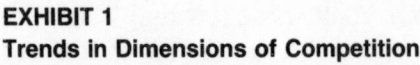

EXHIBIT 1
Trends in Dimensions of Competition

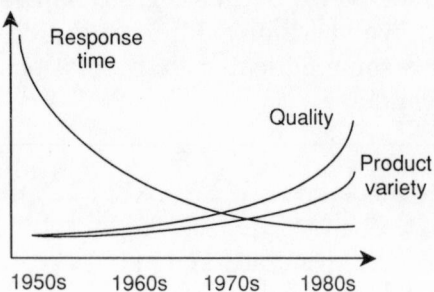

toward rapidly shrinking development times, accelerating rates for new-product introductions, and, consequently, shorter product life cycles.

These trends suggest strongly that time-based competition, like quality, is not a fad. In addition, response time did not suddenly emerge as a factor of competition; it grew steadily in importance. The empirical evidence demonstrates a persistent, unmistakable, downward trend in response time in virtually every industry from World War II to the present day. Will this trend in market competition continue? The answer seems to be obvious. The pattern of competition for the 1990s and beyond can be confidently projected. Unless a fundamental change occurs in consumer behavior, competition will become increasingly time-based.

The trends in Exhibit 1 are driven by forces that can be altered only by a fundamental dislocation in a market-driven economy. Consumers drive the business process instead of the reverse, and the message they send is unambiguous. Through their actions and ad campaigns, producers invite consumers to "Have it your way," and consumers obviously prefer it that way. Why would consumers who have received fast service henceforth settle for slower? Would they develop a preference for shoddier quality? Less variety? Barring restrictions on competition, market response time will continue to become faster, consumer expectations for product and service quality will continue to escalate, and product and service variety will continue to expand. Customers want it this way and their expectations are constantly ratcheting upward. Competition also drives it in that direction.

The single unpredictable factor is the rules of the game. Industry studies tend to confirm certain hypotheses about how time is affecting competition across the entire range of products and services. Research can forecast, confidently, the level of consumers' expectations and the characteristics of the strongest competitors. What analysts cannot do, however, is predict how the playing field for business competition will be altered by environmental and political forces. The situation is analogous to predicting the future of competition in a sport such as basketball. With basketball, few can forecast how governing bodies, such as the NCAA or NBA, will tinker with the rules, but experts agree that the most competitive teams will have players who have exceptional mobility, handle the ball more adroitly, jump higher, and shoot accurately. These teams should thrive no matter how the rules change.

The same proposition is true in manufacturing. The time-based competitor should thrive, because by achieving a time-compressed process, the firm gains the capability to excel on all the other important dimensions. So, no matter what trade restrictions are enacted or what environmental constraints are imposed, the time-based competitor should have a leg up on the competition. While one may not be able to predict the rules, one can describe the skills of the most successful players. The time-based competitor will have the most powerful competitive advantage in the marketplace ahead.

Furthermore, it can be stated with confidence that competition will intensify along these dimensions. In many industries, high quality is assumed and, therefore, among those firms with world-class quality, time is emerging as the *distinguishing* dimension of competition—that is, the one dimension that separates the leader from the pack.

INDUSTRIES REACTING SLOWLY

The shock effect of an emergent time-based competitor will be most pronounced in those industries that have been slow to develop quick response. A specific example is the plastic injection mold industry. In 1989, field research was undertaken to determine how mold builders and their customers, molders of plastic parts for end-users, valued response time. Would a molder pay a substantial price premium to the mold builder who offered a shorter lead time? Since time had been found to be of growing competitive value in other industries, the responses from the mold builders first surveyed in the research were shocking. They expressed the view that time is not an important factor of competition in their industry. The molders of plastic parts stated that, in selecting a mold builder, high quality was assumed but that builders do not compete against each other on the basis of time. A molding plant owner stated that "lead times in our industry average 14 weeks. That's fine, we can factor that into the schedule, but cannot tolerate a late shipment. If the builder says 14 weeks and is 2 weeks late, I'll never deal with that builder again." Consistency of lead time is valued more highly than reduced lead time.[1]

This initial research indicated that the injection mold industry was a rarity: an industry that was immune to time-based competition. When further research inquired as to how much molding firms would pay to receive faster delivery of molds they ordered built, their answer was "very little." The typical mold builder is a small, privately held machine shop with fewer than 20 employees. Many of these shops do appear to compete for small jobs based on price. In fact, competitive bidding is often so fierce that the outcome is a "winner's curse"—the one who wins the bid has made a mistake in costing out the job and is guaranteed to lose money on the deal.

Further study of the industry revealed that, to the contrary, mold builders were not only quite vulnerable to time-based competitors, but they were also about to be blind-sided by this new form of competition. The pressure for time compression in this industry is coming from sophisticated end-users of plastic parts,

---

[1]Anthony J. Zahorik, Timothy L. Keiningham, and Joseph D. Blackburn, "Time-Based Competition in the American Plastic Injection Mold Industry," Operations Roundtable Research Report, Owen Graduate School of Management, Vanderbilt University, November 1989.

such as IBM, Xerox, and Motorola. As global competitors, these firms demand shorter tooling times in their development cycles and actively seek strategic alliances with time-focused mold builders. According to Bob Vines, early manufacturing involvement manager at IBM in Lexington, Kentucky, "Time is everything. It's well worth it to pay double to cut production time in half."[2]

End-users such as IBM put a premium on time, and this makes these firms a catalyst for change in the mold industry. They are seeking out and directing their business to those mold builders who can deliver high-quality molds with the shortest lead time. Minco Tool of Dayton, Ohio, is typical of mold-building firms that are reaping the benefits of orders from the time-based end-users. Minco charges a price premium of 20 to 30 percent for the fast delivery of high-quality molds that are guaranteed to work the first time (and for a manufacturer, the lead time that counts most is the time that elapses from placement of the order to production of the first good part). By emerging as a time-based competitor, Minco has become one of the largest and most successful mold shops in the United States.

Other firms like Minco compete on the basis of response time and, consequently, boast extremely profitable businesses because they can command a substantial price premium for fast service. These firms are a small, but growing, segment of an industry that is on the brink of upheaval. Today, the time-based mold builders are outnumbered by the small, business-as-usual mold builders. Earlier research missed the change occurring in mold building because the initial survey sample was so heavily populated with these conventional mold builders. Hindsight discloses that the researchers "talked to the wrong people."

The small, conventional mold builder is in grave danger. The barbarians have entered the gates, and yet these builders do not recognize the enemy. The time-based competitors are in place and beginning to flex their muscles by skimming off the cream of the business and leaving the builders who lag behind to compete for a smaller, less profitable share of the business.

---

[2]Ibid.

TIME IS NOT A TRADE-OFF

Many U.S. managers are wary of adopting response time as a strategic paradigm because they fear that, if they do, quality or cost will suffer. Conventional wisdom warns that for a company to gain improved response time, it must give something else up, such as quality, cost, or flexibility. The mindset that views all strategic decisions in terms of trade-offs has been ingrained in an entire generation of managers, mainly, by business schools.

Focusing on response time as a strategic weapon, however, is not a zero-sum game. In hundreds of companies, the JIT experience shattered the conventional beliefs about trade-offs: as a JIT process was implemented, the companies' response time and quality improved in parallel. Faster response time does not cost more; it costs less. A faster process tends to be simpler and, as a result, more flexible. *The power of a time-based strategy is that, by focusing on speed, the firm develops world-class quality and the process flexibility to deliver a wider variety of products and services without the burden of increased cost.*

The trade-off mindset among U.S. managers prevents many firms from moving to become time-based competitors. The barrier is that the recipe sounds "too good to be true" and flies in the face of long-held beliefs—all those things learned in business school. Virtually all the early evidence that a response-time strategy did not involve conventional trade-offs was provided by the examples of Japanese firms. So long as these firms operated solely in Japan, American managers found it easy to dismiss the evidence and to explain the results in terms of cultural differences. Now that a substantial number of major U.S. firms have developed response-time capability and, consequently, demonstrated that the process of change is largely culture-free, the evidence is too strong and persuasive to ignore. Firms that cling to the conventional view that strategic choices must be made among time, quality, price, and variety will eventually learn differently from the time-based competitors within their industries, but they may learn too late to survive.

MASTERING THE PROCESS

Because response time likely will play an increasingly weighty role in determining competitive advantage in the 1990s, firms that understand and apply the *process* of time compression can be expected to gain an upper hand within their markets. Who will these firms be? The writer's research leads to the conclusion that the ascendant manufacturing firms will be those that do two things: first, develop JIT manufacturing; second, extend JIT concepts throughout other functions of the organization. The recurring theme of this book is that JIT is the model for taking time out of all parts of the value chain. Just-in-Time is essentially a prescriptive model for removing all wasted time from a process, simplifying it, and making it flexible. American managements know this process, and many U.S. firms have become adept at implementing it; it is not some secret process, shrouded by the trappings of Japanese culture.

In their rush to emulate the Japanese, however, U.S. firms have focused too narrowly on the manufacturing cycle. In most industries where the time consumed by all activities in the value-delivery chain has been examined, the analysts discovered that manufacturing consumes a small fraction, often under 5 percent (and shrinking), of the total time required from raw materials to delivery of a product to customers. To the extent that U.S. industry has attempted to adopt JIT, it has focused those efforts exclusively on shortening the manufacturing cycle. What is U.S. industry doing to shrink the remaining 95 percent of the time in the pipeline? An American firm that endeavors to become a global time-based competitor will find that it is not enough to simply pursue JIT to achieve short-cycle manufacturing (even though that alone is a daunting challenge for many firms today). The reason is that the manufacturing function is a diminishing fraction of the total time in the value-delivery pipeline that must be compressed.

Firms that can successfully implement JIT production, however, will gain a huge advantage over those that do not because the JIT lessons learned will sharpen their skills for time-compressing other operations. Such firms will develop the *process* that will enable them to eliminate waste from the remaining 95 percent of time spent on nonmanufacturing activities. As shown in preceding chapters, functions outside the factory walls—new-product development, customer service, distribution—can benefit from applying the JIT model. These other segments of the chain, moreover, present greater windows of opportunity than the manufacturing cycle.

To focus on time in the entire value-delivery cycle, management must adopt response time as an organizational imperative; it must commit to becoming a time-based competitor, and it must communicate that commitment throughout the organization. The firms that succeed will do so because they have the clarity of vision to see the general applicability of the JIT experience. The "learning organization" will seek to propagate throughout the firm the technology of time compression that is the essence of JIT.

WHITE-COLLAR JUST-IN-TIME

Since many firms have achieved short manufacturing cycle times, the greatest untapped opportunity for time compression exists in the office. In manufacturing, the volume of information flows and the paperwork often far exceed the product flow. A growing fraction of the total time in the value chain is devoted to information processing—customer service, order processing, credit approvals, engineering design changes, and so forth. This is the "hidden factory" that accounts for the rapid expansion of overhead costs in manufacturing.[3]

The challenge for the time-based competitor is to manage these white-collar information processes as efficiently as the firm manages the blue-collar production processes. If the information-processing operations in offices are studied and mapped in the same way as flow diagrams for manufacturing, the two operations look similar. In the white-collar environment, one finds time-consuming setups, batch processing, and sequential activities. The net result is the same as in manufacturing—delayed response to the market, higher cost, and poorer quality.

---

[3]Jeff Miller and Tom Vollman, "The Hidden Factory," *Harvard Business Review*, Sept.-Oct., 1989, pp. 142–150.

Although much more research is needed, the early returns indicate that the concepts of JIT production can be no less powerful when applied to white-collar activities. In this environment, any distinction between manufacturing and service industries disappears. In fact, many of the leading time-based competitors are service firms, such as Citicorp and Federal Express, that now successfully apply these concepts to speed up information processing.

One disturbing sign indicates, however, that many firms are repeating, in the office, a pattern of mistakes that they made on the shop floor. The term "office automation" describes the approach by which firms seek to use technology—computers, lasers, and telecommunications equipment—to increase productivity in office processes. This type of strategy was followed by the factory automation advocates who believed that robotics, automatic-guided vehicles, and laser cutters could be inserted on the factory floor and, thereby, leapfrog in efficiency the Japanese manufacturing processes. The shortcoming of the "automation fix" is that no fundamental changes are made in the operations sequence and, consequently, an inefficient, antiquated process is merely automated. Attacking white-collar productivity with automation is a superficial approach that gold-plates the deep-seated problems that curtail that productivity.

Many manufacturing firms are just now beginning to see the untapped potential for time compression in their white-collar functions. Among these, a smaller group of firms will recognize the relevance of JIT techniques for other functions, such as new-product development and customer service. The truly successful time-based competitors of the 1990s will be the ones who see the problems in their white-collar functions and transfer the technology from their manufacturing conversion to solve those problems.

A FUTURE AGENDA

A lot is yet to be learned about time-based competition. Research has begun to grasp its power as a strategic weapon, but it has just scratched the surface on how to forge organizations into deployers of this type of weapon. In the white-collar area, for example, only shreds of evidence exist to support the hypothesis that JIT concepts can be effective in removing wasted time. These data are mostly anecdotal, since most of the experience with JIT has been on the shop floor. Few firms have reported experiences in applying JIT to new-product development, and, in fiercely competitive industries, successful firms have a strong incentive to keep their results secret.

Likewise, much remains to be learned about managing a time-focused corporation. It is clear what must be done to achieve time compression, and the benefits are recognized. However, far too little is known about how to effect change and to manage the implementation process. To be an effective time-based competitor, a firm must seek to remove time from *all* segments of the delivery chain. Since this requires the joint efforts of all functional groups in the organization, it is unlikely to be a process driven by a single function (such as manufacturing). Motivation and commitment must come from the top. In fact, the leading time-based competitors in the United States—Hewlett-Packard, Northern Telecom, Federal Express—have that kind of leadership. In their messages to the public and to their employees, the CEOs of these firms now use the term time-based competitor to characterize their firms' strategic focus. Gaining the commitment of an entire organization to a strategy objective such as time raises issues of leadership and motivation that have yet to be solved. In Chapter 12, Dan Ciampa of Rath & Strong, one of the experts on the subject of leadership for change at the CEO level, eloquently expressed his thinking on the subject but, at the same time, indicated that this crucial area needs much more study.

A few things are well-known about the organizational structure of the time-based competitor. The traditional corporation is organized by function; engineers work and talk to engineers and manufacturing is in a separate building or even in another part of the country. However, the manufacture of a product or delivery of a service is a multifunctional effort. Taking time out of this process requires coordination across functions. The suspicion is growing that, to be most effective in compressing time, the firm needs to be organized around processes, rather than functions. One hypothesis that this suggests is that the time-focused firm should be organized around teams. If so, what are the implications for our traditional standards of performance measurement, promotion, and so forth?

Many firms feel that a team approach is more effective for time compression than a traditional functional organization. As John Bailey described in Chapter 6, teams have proved to be quite effective in cross-functional activities such as new-product development. Much remains to be learned about team construction and motivation, not only in new-product development but also in manufacturing, customer service, and other parts of the pipeline. The question of whether teams can be more effective in white-collar activities, to my knowledge, has not been addressed in a systematic way.

To summarize, a growing body of evidence is asserting that time will be an increasingly important strategic weapon. Firms are learning what it takes to become time-based competitors, and it is not something that can be purchased from a supplier or uncrated or installed. Time-based competition requires a fundamental reconstruction of the processes by which goods are manufactured and services are delivered. More than that, it may require rethinking the way those processes are managed.

The firm of the future must deliver on all three dimensions of variety, response time, and quality, so the firms that prosper will be those that assemble the entire puzzle. The theme of this book is that a time focus is the way to make simultaneous progress on the other dimensions. The firm with a future must concentrate on the entire value-delivery chain—shorter manufacturing cycles are not enough. Moreover, the manufacturer who develops a time-compressed delivery chain will possess a huge, exploitable global advantage because the process is transferable into new countries, new markets.

The challenge to the firm of the present is inescapable. The customer is demanding it all—faster response, more variety, newer product features, higher quality—at a competitive price. The firm of today must decide either to become a time-based competitor and meet these needs or retreat from the business. In many industries, competition has driven quality to such heights that it is no longer a distinguishing factor: Every significant player in the market exceeds the quality standards required for admission to the game. With high quality as a standard, response time often becomes the differentiable factor among competing firms. For instance, the tide is moving in that direction in plastic injection mold building, and those firms unaware of the flow will soon be swept away.

In any industry, firms must move quickly; the process of becoming a time-based competitor takes time; no shortcuts are apparent. Toyota Motor Company showed that a time-focused process is developed through millions of incremental steps—the process of Kaizen. To persist in this process requires the patience embodied in the old Chinese proverb, "A journey of a thousand miles begins with a single step." It requires a long march, but one that cannot be delayed because each day of delay forestalls progress vis-à-vis the competition. Toyota perfected the process, which is now the model for all time-based competitors, over several decades. Toyota was able to develop its weapon patiently and in relative secrecy. Today the secret is out; firms must deploy their own speed weapon or plan to surrender to their time-based competitors. Delay means destruction.

INDEX